DISCOVERING ECOLOGY

Opposite: This African leopard (Panthera pardus), *camouflaged from its prey by its spotted fur, is resting in the tree-tops.*

DISCOVERING ECOLOGY

by Tim Shreeve

Longman

Longman Group Limited
Longman House
Burnt Mill, Harlow, Essex, UK

First published in Great Britain 1983

Created, designed and produced by
Sceptre Books Ltd., London

© Sceptre Books Limited 1982
All rights reserved. No part of this book may be
reproduced or utilized in any form or by any means,
electronic or mechanical, including photocopying,
recording or by any information storage or retrieval
system, without permission in writing from Sceptre
Books Limited, London, England.

Set in Monophoto Rockwell Light by
SX Composing Limited, Rayleigh, Essex.
Separation by Adroit Ltd., Birmingham
Made and printed in Spain by Novograph S.A.

British Library Cataloguing in Publication Data
Shreeve, Tim
    Discovering ecology.
    1. Ecology—Juvenile literature
    I. Title
    574.5   QH541.14

ISBN 0-582-39220-9

Contents

The Scope of Ecology	6
Systems of Organization	8
The Soil	10
Soil Types	12
Photosynthesis	14
Nutrient Cycles	16
The Transfer of Energy by Plants	18
The Ecological Niche	20
Communities	22
Succession	24
Climax Communities	26
The Ecosystem	28
Energy in Ecosystems	30
Ecological Pyramids	32
How Nature Controls Populations	34
Competition and Coexistence	36
Close Relationships in Nature	38
The Diversity of the Earth	40
Freshwater Ecosystems	42
Marine Ecosystems	44
Life in Estuaries	46
Coral Reefs	48
Mangrove Swamps	50
Tundras	52
Deserts	54
Grasslands	56
Forests	58
Coniferous and Deciduous Forests	60
Tropical Rain Forests	62
Ecology and People	64
Soil Erosion and Conservation	66
Mankind and the Forests	68
Mankind and the Seas	70
Conserving Wildlife	72
Desert Increase	74
The Blooming of the Deserts	76
The Growth of Human Populations	78
The Control of Diseases	80
Pollution	82
The Problems of Intensive Agriculture	84
Biological Pest Control	86
The Great Experiment	88
Food for the Future	90
Recycling Resources	92
Glossary	94
Index and Credits	96

The World of Ecology

Life exists almost everywhere on Earth. Wherever we live we are surrounded by rich and beautiful patterns in nature, both on land and in the sea. Even the hottest deserts contain a few plants, such as cacti, and animals such as snakes and insects. Some plants and animals, including polar bears and penguins, can also live in the very cold regions near the North and South Poles.

Cities and towns may appear lifeless compared with the countryside. If we look closely, however, we can usually find patches of wasteland, as well as parks, that are full of trees, flowers and animals. Buildings are often homes for a variety of insects and sometimes small rodents. Such inhabitants are not, of course, always very welcome!

Unfortunately many of us tend to ignore our natural surroundings and the plants and animals that live in them, failing to realize how important nature is to human beings. We rarely consider, for example, what happens in the soil beneath our feet. The soil, in fact, teems with life, containing millions of tiny creatures that break down plant and animal remains into food for living plants. Similarly, although we may pause to admire a lovely landscape, how often do we stop to wonder why plants have green leaves and grow best where there is plenty of sunlight, why the rain falls or where it comes from and why the plants and animals of rivers, oceans, forests and fields are so different from each other? Ecologists ask themselves questions like these and try to find the answers to them.

Discovering how nature works is the basis of ecology, and ecologists study nature by looking at how all the living and non-living things on Earth relate to each other. Plants and animals, for example, are linked by their feeding habits. Plants are eaten by certain kinds of animals – and some animals will eat only particular plants. Plant eaters are eaten by meat-eating animals that are frequently, in turn, preyed upon by others.

All plants and animals also have to adapt to their surroundings to make the best use of them. Trees release water vapour through their leaves so, in areas with cold winters when there is less water available, some trees shed their leaves and thus conserve water to stay alive. Some animals, such as hedgehogs, hibernate during cold periods when it is difficult to find food.

Conditions are, however, constantly changing. These continual alterations have caused new types of plants and animals to evolve while others have become extinct. This is the reason for the immense variety of different plants and animals all over the world today.

People have come to dominate the Earth and, in so doing, have made great demands on nature. We have used plants and animals for food and trees for building materials; we have cleared vast areas of vegetation and killed many animals for their skins or because we regard them as pests. These activities are dangerous, for they disturb the delicate balance of nature.

Applied ecology teaches how we, too, are part of the world around us and how we can improve our planet rather than destroy it. The world of ecology covers many aspects of nature and is not only a fascinating subject to learn about but is also of vital importance for our future on Earth.

The Scope of Ecology

Ecology is the study of organisms – living things – and how they exist in their environment. The environment of any organism consists of all the living and non-living things on which it depends for its existence together with the factors that control its lifespan. The environment of a plant, for example, consists partly of the soil, the water and food contained in the soil and the atmosphere in which the plant is growing. The plant also depends on the climate and the weather – forces that may determine whether it can live and reproduce. Other organisms may also exist alongside the plant – these could be grazing animals that may eat the plant, and other plants that may compete with it for food and space in which to grow.

There are two ways of studying ecology. The ecologist can consider a single species and find out how it lives in its environment. This method is known as autecology. Alternatively, the ecologist can look at how all the different organisms in a particular place relate to one another and to their non-living environment. This second method is known as synecology, or the ecosystem approach. Both methods are interesting ways of studying the natural world, and involve examining events that change with time.

Although organisms affect their environment, they are, at the same time, affected by it themselves. A single bee, for example, flying between flowers searching for pollen and nectar, fertilizes the flowers that it visits. Most bees are selective in the flowers to which they go, preferring some species to others. If the bee is the only pollinator in a single area, and all the flowers must be pollinated to produce seeds, then only those flowers that the bee visits can reproduce. In succeeding years more flowers would be available for the bee to exploit, as those that it prefers would reproduce, replacing flowers that were not pollinated.

A single bee, however, is rarely the sole pollinator in any region. Other insects, such as butterflies and hoverflies, also visit and pollinate flowers, often different from those the bee visits. Therefore, although the bee is gradually changing its surroundings by its pollinating activities, other insects are also altering the environment of the bee.

Ecology enables us to understand the far-reaching

A honeybee is seen here gathering nectar and pollen from heather. As the insect moves between the flowers, it transfers pollen and so causes the flowers to be fertilized.

effects that human activities, such as farming and building dams, have on the environment. Ecology can also help us to solve puzzles – for example, why zebras are found in African grasslands but not in adjacent forests, why tropical soils often produce poor crops after a few years of cultivation, and why Antarctic penguins contain high levels of the insecticide DDT despite the fact that DDT has never been used in the Antarctic.

In order to answer the first question the ecologist investigates how the zebra has adapted to grasslands and looks at the zebra's relationships with its food and predators. The ecologist must also compare the zebra's grassland habitat, or living space, with the forest habitat. To find out why tropical soils often fail under cultivation the ecologist has to consider how nutrients are cycled between plants and the soil, and find out the reasons why a forest is able to grow in a tropical soil but crops fail in the same soil. To discover the relationship between penguins and DDT the ecologist must learn how DDT travels from the land into the sea and follow its path through a chain of marine organisms until it reaches the penguin. Ultimately, it is the goal of the ecologist to find out if events such as these, that may seem disconnected or even accidental, fit into a pattern that will show people how to understand and live in harmony with the workings of the Earth.

These Emperor penguins breed only on the most southerly coasts of Antarctica, on floating ice. The female lays her single egg during the bitter Antarctic winter, and then passes it to the male who carries it on his feet, keeping the egg warm by draping a fold of skin over it. Six to eight months later, the chicks have developed and can feed on the plentiful food of the short summer. Unfortunately, DDT, absorbed by marine organisms, which are eaten by fish and eaten in turn by penguins, has produced high levels of the insecticide in penguins and their chicks.

Systems of Organization

The basic living unit with which ecology is concerned is the individual. Individuals differ in their composition. Organisms such as bacteria, some algae and protozoans (amoeba, for example) consist of a single cell. But almost all organisms people encounter in daily life are multicellular. Most familiar plants and animals are recognizable as single individuals. Occasionally, however, it is difficult to tell where one individual ends and another begins. With some types of coral, for example, the individual coral animals, or polyps, are fused together and share a connecting skeleton. So a colony of such corals may be regarded as a single individual for most of its needs, including feeding – but each coral polyp reproduces on its own.

Insects that live in colonies, such as termites and honeybees, are easily identifiable as individuals, although each one has a role that benefits the colony as a whole. They are often called social insects because of this special behaviour. Usually only one member of the colony, such as a queen bee, is able to reproduce while the majority feed and defend the colony. Colonies of these social insects are sometimes referred to as super-organisms since the entire colony functions almost as an individual.

Individuals of the same species form units that are

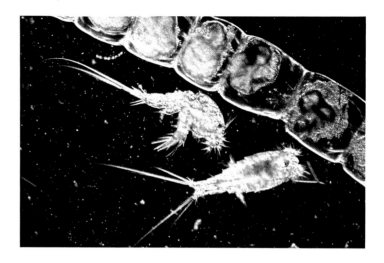

A sea community of green algae and other plants (phytoplankton) and minute animals (zooplankton) is seen here under the microscope.

Individuals within a particular habitat form groups of the same species called populations. These populations interact with populations of other species to form a community. Communities react to physical forces such as water, air and sunlight to form an ecosystem.

individuals

populations

called populations. Populations are found in distinct areas, and a single species usually consists of several population units, each in a separate place. Some populations may contain individuals tightly packed into a small area, such as whitefly on cabbage plants. Other populations, such as buzzards on mountains, may be extremely scattered with large distances between each individual. Both types of population work efficiently because the individuals are able to meet others of their kind and breed.

Most areas contain populations of more than one species. Even a wheat field under the most intensive agricultural production will contain populations of different species of micro-organisms in the soil and, in all probability, a population of aphids on the wheat plants themselves. This assemblage of populations in an area is known as a community.

Within a community there are always interactions between the various populations. Some may be competitive and others may be cooperative. Many species of ants, for example, have colonies of aphids. Aphids eat plant sap and secrete a sugary solution that ants find attractive and feed to their young. Ants will defend their aphid farms from attack by parasites and other predators, such as ladybird larvae. The ant and the aphid therefore have a cooperative relationship. The aphid gains protection, and the ant obtains sugar from the aphid. The relationship between the ants and the aphids' predators is competitive. Both groups of animal compete for the aphids.

These ants, on a golden willow in Ithaca, New York, are 'tending' aphids. The ant and aphid populations have a cooperative relationship.

Communities react to the abiotic or non-living environment. That is the part of the environment, including air, water and minerals, that is incapable of life processes such as growth and reproduction. The community, together with its non-living environment, is known as an ecosystem. Ecosystems are dynamic. Not only do their populations develop into new forms and replace each other in time but the non-living environments themselves are also constantly changing – because of the movement of water and wind, for example, or the varying seasonal strength of sunlight.

The Soil

Soil is made up of two parts, a non-living mineral component and an organic component consisting of living organisms and their waste products. The majority of soil organisms are microscopic bacteria, fungi and algae. Other common organisms are larger, such as centipedes, mites, beetles and earthworms. The living part of the soil gives it fertility and determines whether plants can grow in it. Soils that are composed mainly of mineral particles, such as salt flats, support few or no plants. Other soils that contain many living organisms, such as soils under forests and grasslands, support plentiful vegetation.

Soils are created by the weathering of rocks and the activity of micro-organisms. Rock is broken down into small particles by the action of water, wind and land movements. The first organisms to colonize this weathered rock are often tiny blue-green algae (so-named because of their usual colouration – although some may be red) and lichens. Both these plants convert nitrogen from the atmosphere into nitrates, and convert carbon dioxide and water into complex starches. The presence of these nutrients and the blue-green algae and lichens themselves enables other plants and simple kinds of animals to move into the developing soil. These invading organisms are the fungi, bacteria and single-celled animals that feed on dead plant and animal remains. The feeding activity of such organisms, known as decomposers, frees nutrients such as nitrates, phosphates and sulphates into the soil. These nutrients can be absorbed from water in the soil by the roots of higher plants, such as grasses, that colonize the new soil.

The waste products of the decomposers are a mixture of very complex chemicals called humus. These are sticky compounds that give the soil texture by binding minerals into small crumbly particles. The texture of the soil is important because there are spaces between the soil crumbs that are filled by gases and water. The size of these spaces determines how quickly water and plant nutrients are washed from the soil by rain-water that trickles through it. The air spaces also provide oxygen, nitrogen and carbon dioxide for plant roots and micro-organisms.

Organisms that move in the soil, such as earthworms

This deep section of soil in Iraq clearly shows how an alluvial soil – one that is washed down by rain or rivers – has been built up in layers. The different layers originate from different areas.

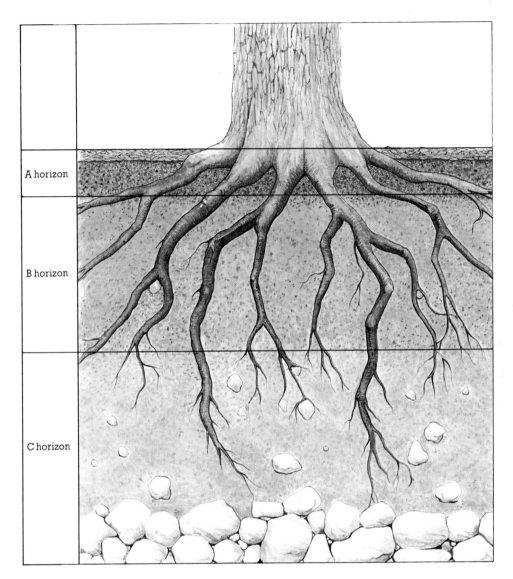

This earthworm is shown with its cast, which consists of undigested soil that has passed through the worm's body and is excreted in coils, and left on the soil surface.

The diagram on the left shows the three soil horizons typically found in soil in a temperate climate.

and beetles, are also important for soil development. They carry the dead plant and animal remains from the surface into the lower layers of the soil and break up large particles into smaller pieces. This action allows the decomposers to break down the remains more quickly.

A trench cut into soil usually reveals distinct soil layers that are often of different colours and textures. These layers are known as horizons. The top, or A, horizon is composed of dead plants and animals that are being broken down by soil animals and digested by the decomposers. At the bottom of the A horizon there is often a layer of humus that has been deposited by rain-water. The B horizon is composed of mineral soil containing the nutrients produced in the A horizon together with the products of the chemical and biological breakdown of humus. It is the top two soil horizons that provide growing plants with most of their nutrients. The bottom, or C, soil horizon is composed mainly of so-called parent material, the original substance from which the mineral part of the soil was created. The parent material may be bedrock, millions of years old, or may consist of material that has come from other regions.

Soils that have been moved by glaciers, wind and water are often extremely fertile because they may consist entirely of topsoil from other areas. The Ganges delta, for example, is composed of soils that have originated in the Himalayas and the plains of India, and have been deposited by the Ganges River at its mouth.

Soil Types

Different areas of the world have different types of soil on which specific kinds of vegetation grow. Soils in the hot tropics and subtropics, those regions near or on the Equator, are varied, ranging from desert soils to highly fertile soils, sometimes derived from volcanic lava washed down to plains by rain or in rivers. In temperate regions there is a group of fertile soils known as brown earths because of their colouring. These soils are found beneath grasslands and woodlands where the trees shed their leaves in winter. In colder areas of the world, where coniferous forests grow, the soils are less fertile. In the Arctic and some parts of the subarctic there is permafrost, a layer of ground below the surface which is frozen all the time. Nothing can grow in frozen ground, but when the thin surface layer above the permafrost thaws during summer, hardy plants such as sedges, mosses and short grasses can briefly flourish.

Within any single area there are often local variations in soil type. The most important factors that influence soil development are temperature, rainfall and evaporation, and the nature of the material from which the soil is derived. Vegetation cover – trees, shrubs, grasses for example – also has an effect, although it is influenced itself by the kind of soil in which it is growing.

High temperatures in the tropics, combined with sufficient water, make soil organisms active. Decomposition is therefore rapid. The rainfall that penetrates the soil is also warm and encourages rapid decomposition. The warm water dissoves the soil mineral silica and this, together with humus, enables the soil to hold plant nutrients that are soluble, meaning that they can be dissolved in water. The loss of silica increases the proportion of iron and aluminium salts in the upper soil surface, often giving tropical soils a reddish colour. This process, involving the loss of silica from the soil, is known as laterization.

Many of the arid soils of tropical and subtropical regions are characterized by the accumulation of calcium and sodium salts on, or just under, the soil surface. These salts have been carried upwards by the evaporation of water from the surface of the soil. Such calcified and salinized soils are often found in areas that are irrigated or have high ground water levels. They support little vegetation because water in the plants is drawn from the roots by the salt concentration in the soil.

In temperate regions the cool rainfall does not dissolve silica. Brown earths are often very stable because of the even flow of water from light but frequent rainfall. In brown earths there is little upward movement of water and dissolved salts, so the problems of laterization, salinization and calcification, found in the tropical and subtropical soils, do not occur.

The soils under coniferous forests are acidic because coniferous plant material decomposes into organic acids. These acids dissolve clay particles on the surface and the downward movement of water deposits the clay deeper in the soil. This process, known as podsolization, reduces soil fertility because nutrients are moved downwards with the clay beyond the depths to which roots grow. The soil is also too acidic for earthworms which, in brown earths, help to mix the plant litter into the mineral soil.

In tundra regions, such as this one in Alaska, the movement of thawed water through the ice of the permafrost soil causes these polygonal wedges of land to be formed.

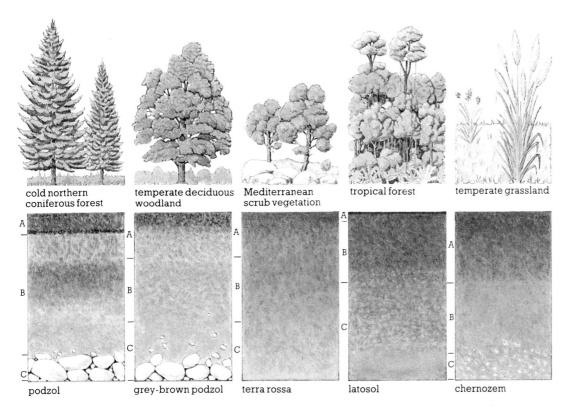

This diagram shows five typical soils: a podzol from northern France; a grey-brown podzolic soil from southern England with a richer, deeper A horizon supporting deciduous woodland; a terra rossa (red earth) from Portugal; a latosol, or tropical laterite soil, from Puerto Rico in Central America; and a prairie soil, called a chernozem, from South Dakota, U.S.A.

These white sands in New Mexico, U.S.A., are formed by pure calcium sulphate in which plants cannot grow, despite the suitable climate.

Photosynthesis

A green plant can grow if it is provided with sunlight, water, air and minerals. Plants achieve growth by a series of complex chemical reactions, called photosynthesis, that involves trapping the energy of sunlight. The rate at which photosynthesis takes place depends on the amount of light, water, and minerals that plants receive, and the temperature of their surroundings.

The green colouring matter in plants is called chlorophyll. Chlorophyll can absorb sunlight. Once the Sun's energy is trapped inside the plant, the chlorophyll generates tiny electric currents. This electricity is used to split water, which the plant's roots have taken in from the soil, into its component elements of hydrogen and oxygen.

Carbon dioxide and oxygen in the air are also absorbed by plants. These gases can enter, and leave, the plant by means of small openings, called stomata, in the cuticle, or skin, of the plant. Within the plant, the carbon dioxide and hydrogen are converted, by a complicated chemical process, into sugar. Water is also formed as a by-product. Some of the sugar is stored in the leaves as starch and some is combined with minerals, such as potassium, phosphates and sulphur, to create the compounds that are used to build the plant tissues of stems, leaves and flowers.

Some of the water that is formed by the conversion of hydrogen and carbon dioxide, and some of the water that is absorbed by the roots of the plant, evaporates from the leaves. The oxygen that is produced during photosynthesis is released into the air through the

These diagrams show the process of photosynthesis, which enables green plants to convert light energy into the chemical energy of sugars, and the complementary process of respiration, which releases energy. Both processes use enzymes, proteins that promote chemical change.

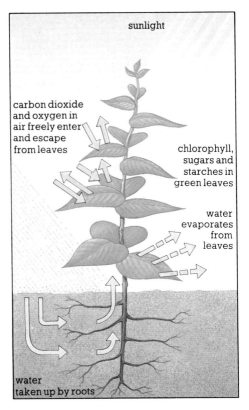

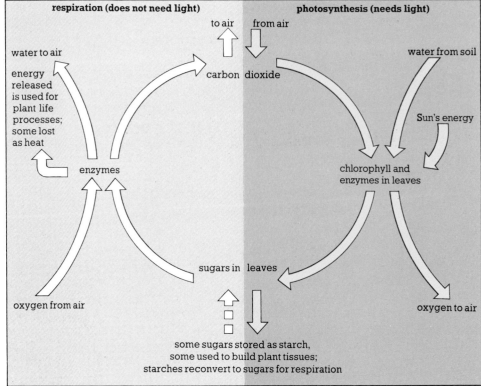

Sugar cane grows quickly in warm places, such as Cuba, where it is an important crop for export. Cane is harvested by cutting down the plant stalks which are then pressed to extract the juice. The concentrated juice produces a brown, sticky sugar. Refined sugar, which has the non-sugar parts removed, is less nourishing than unrefined sugar.

stomata. Green plants are, in fact, the source for most of the oxygen in the atmosphere of the Earth.

Photosynthesis can take place only during the daytime when there is plenty of light, when temperatures are higher than at night, and when there is less water in the atmosphere. These conditions cause the stomata to open and allow water to evaporate through them. In dry conditions plants without special adaptations often wilt during the day because they lose more water than they can replace by drawing it up from the soil.

Plants that are normally found in arid conditions have special mechanisms to overcome this problem. Some, including desert cacti, reduce their water loss by opening their stomata only at night. They release oxygen and absorb carbon dioxide at the same time. The carbon dioxide is stored for use in the daytime by being converted into an acid, which takes up less space than a gas. Cacti and other similar plants grow slowly because most of the sugar produced during the day is used in the storage and release of carbon dioxide.

Other types of plants, such as maize and sugar cane, take in carbon dioxide during the day. The water produced by photosynthesis during the formation of sugar is re-used by the plant instead of evaporating through the stomata. In high temperatures (over 25°C) these plants grow quickly.

In the oceans plant growth is not limited by water or light but by lack of minerals – usually phosphates – as these tend to sink to the bottom. In shallow estuaries and coral reefs minerals are abundant and the amount of plant growth – often algae – is equal to that in tropical forests. In swamps and marshes, which are shallow, warm and moist, there are large supplies of minerals, very high light intensities and ample amounts of water. In these areas plant growth is very rapid.

This cactus tree withstands desert conditions because it opens its stomata only at night when the humidity of the air is higher than in daytime, thus reducing water loss. It grows very slowly and has sharp thorns, which discourages animals from eating it.

Nutrient Cycles

Plants and animals require a variety of chemical elements and compounds in order to live and grow. The most important of these are water, oxygen, carbon, nitrogen, phosphorus, sulphur and calcium. Others, known as trace elements, such as the metal iron, are needed in smaller quantities but are still essential for life.

In the non-living environment, elements are present in the form of gases and vapours (such as oxygen, nitrogen and carbon dioxide) in the atmosphere, and as dissolved and solid chemical compounds in water and soil. In organisms, such elements are incorporated into their body tissues. Plants obtain their requirements for these elements directly from the non-living environment. Animals meet most of their needs by eating plants or other animals.

Since living things, for the most part, inhabit a narrow layer of the Earth on or near the surface, only a small proportion of the planet's substances are readily available to plants and animals. Life would quickly end if vital elements and compounds were used only once, passing from the soil and atmosphere to plants and

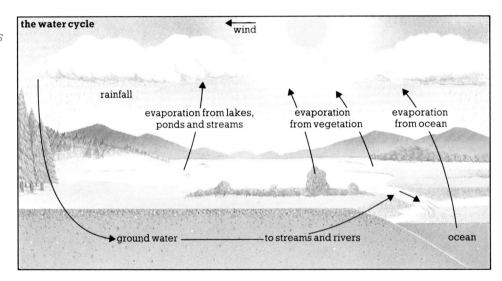

This diagram shows the water cycle. Water falls on to the Earth as rain or snow. Some water drains into lakes and is absorbed by the soil. Water re-enters the atmosphere by evaporation from plants, the soil or from areas of water.

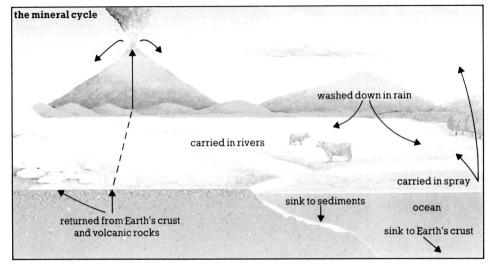

Minerals are washed down onto the land and either absorbed by the soil or drained into rivers, finally reaching the seas and oceans. In the sea, minerals sink to the floor and form sediments. Minerals are returned to the atmosphere in sea spray or through volcanic eruptions, so completing the cycle.

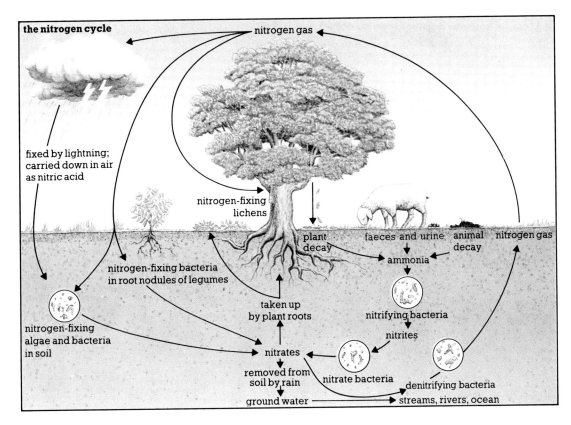

All plants and animals need nitrogen to build proteins for growth, but the nitrogen gas has to be converted first into nitrites, then into nitrates, as shown in this diagram. Some lichens, certain types of algae and bacteria in the soil and in root nodules are able to do this after receiving nitrogen either directly from the air, from decomposing plants and animals, or animal excreta. Lightning also converts nitrogen into nitric acid, which enters the soil in rainfall.

animals, and then disappeared because they had been used up. To avoid this problem, all of the elements that are used by plants and animals are continually recycled between organisms and the non-living environment.

There are many different nutrient cycles. The carbon cycle, for example, involves only carbon dioxide and the products of photosynthesis. Plants absorb carbon dioxide from the atmosphere during the process of photosynthesis. The carbon is incorporated into plant tissues, such as leaves. Some carbon dioxide is also returned to the atmosphere when plants respire, releasing energy by breaking down the sugars made by photosynthesis into carbon dioxide and water. When animals eat plants the carbon in the plant is then incorporated into animal tissues. When both plants and animals die, their remains are decomposed by bacteria and fungi in the soil. The decomposers release the carbon, by their respiration, back into the soil and atmosphere as carbon dioxide.

Oxygen is the second most abundant element in the atmosphere, and is recycled through both the carbon and water cycles. Most of the oxygen in the atmosphere is produced by green plants during photosynthesis. It is used up again by plants and animals when they break down sugar to obtain energy. The oxygen produced by photosynthesis originates from carbon dioxide and water, while that used in respiration is combined with sugars to produce carbon dioxide and water.

The nitrogen cycle is very complex. About eighty per cent of the Earth's atmosphere is nitrogen but plants cannot use it in this form. The nitrogen gas is converted into nitrates through ammonia and nitrites by a process involving types of bacteria and algae that live mainly in the soil.

Nutrient cycles usually do not vary much, but they can become unbalanced – sometimes as a result of changes in climate. Nutrients may then be deposited in nonliving forms, such as marine sediments. Enormous quantities of carbon, for example, were removed from the carbon cycle during the Carboniferous period, which began 350 million years ago. At this time there was lush vegetation on the Earth and a warm, wet climate. When the plants died, the waterlogged soil sealed in the vegetation away from the oxygen needed for normal decay to form peat. Gradually the layers of peat sank and were compressed to form coal.

The Transfer of Energy by Plants

Plants and animals need energy to move, grow, and for their maintenance. Plants convert light energy from the sun into chemical energy by photosynthesis. All animals obtain their energy requirements from plants either directly by eating plants or indirectly, by eating other animals.

The Earth receives an enormous quantity of energy in the form of sunlight, but very little is used in the photosynthetic process. Between twenty and thirty per cent of the light that strikes the Earth's surface is absorbed by green plants. Most of the remainder is reflected or absorbed by water and bare soil. Only one to two per cent of the energy absorbed by green plants is actually stored in the products of photosynthesis, such as starch. Most of it is lost as heat by evaporation of water from the plant or is directly reflected back into the atmosphere.

The figure of between one and two per cent for photosynthetic efficiency, or transformed energy, is common to all plants. Most food crops appear more productive because people harvest a greater proportion of the total plant than of wild, uncultivated plants. High-yielding crops such as sugar beet and maize are, in fact, no more efficient at transforming the energy of sunlight into chemical energy than the natural vegetation in the same area.

The main reason why plants convert such a small proportion of the energy they receive from the Sun is because almost half the total energy of sunlight is in the form of infra-red light, which plants cannot use. The mechanism of trapping the energy of sunlight relies on the visible light spectrum only, especially red and blue light. Under ideal conditions of temperature and light intensity little energy is lost by reflection and evaporation of water, and very high photosynthetic efficiencies can be obtained. In laboratory conditions algae and green plant leaves can convert as much as thirty-four per cent of light energy into the chemical energy of sugars.

Apart from the inability to trap infra-red light, the most common factor that limits photosynthesis is the availability of carbon dioxide. Carbon dioxide is rare in the atmosphere, making up less than one per cent of the total amount. In enriched carbon dioxide atmospheres, which can be provided in laboratories, photosynthesis becomes more efficient. A doubling of carbon dioxide concentration may also double the amount of light energy that is converted together with the amount of plant material produced. The availability of water, particularly in arid regions, may also limit photosynthesis.

Cultivated commercially throughout the temperate zone, sugar beet provides about one-third of the world's sugar. Nevertheless, sugar beet is no more efficient at transferring the energy of sunlight into chemical energy than the weeds that may be growing near this irrigated beet field.

Intense light, suitable water temperatures, abundant nutrients in the upper layers of water and a good supply of carbon dioxide can cause a massive increase in plant production, such as this natural algal bloom on Crose Mere in Shropshire. The algae photosynthesize more rapidly because of the particular environmental conditions. Often, algal blooms are caused by the addition of fertilizers to lakes and ponds and this can result in an imbalance. If the conditions change, the water may no longer be able to support the dense algae, which then decay. The decomposer bacteria that break down the decaying algae rob the water of oxygen, sometimes to such an extent that fish and other organisms suffocate.

Plant production can be increased by adding fertilizers to land and water. Fertilizers are chemical compounds that dissolve in water to free nutrients, such as nitrates and phosphates, which are needed by plants. If ammonium nitrate, for example, is added to the soil, plant production may increase dramatically. Food crop production is usually increased in this manner. If fertilizers (usually those washed by rain from agricultural land) are added to ponds and lakes, spectacular algal growths, called algal blooms, often result.

The growth of leaves, stems, roots, flowers and seeds is called primary production and it is upon this production that all other life forms depend for their supply of energy. Where production is high, as in the tropics, animal life is abundant – more so than in temperate regions where primary production is less. In places where very few plants grow, as in deserts and ice-covered regions, there are correspondingly few animals. Green plants are therefore essential for all other forms of life on the Earth.

The Ecological Niche

As well as nutrients, every plant and animal needs a suitable place to live. The greater water boatman insect, for example, lives in ponds and lakes filled with vegetation and eats insects and animals of its own size, or smaller. The greater water boatman therefore has specific requirements for the sort of pond in which it lives. The pond must contain plants in which the greater water boatman can hide, and the water must be clear enough for the insect to see its prey. The water must be warm enough for the greater water boatman to be active and must also provide conditions that are suitable for its prey. Not only does the greater water boatman feed on other animals but it is also prey for other creatures, such as fish and birds. The description of all the roles of the greater water boatman – how it lives in the pond, eats and provides food for other animals – is called its ecological niche.

In nature, different animals cannot occupy the same niche at the same time. Different species that appear to live in the same niche will, on close inspection, be seen to inhabit different ones. Off the coast of Britain, for example, there are two very similar species of cormorant: the common cormorant and the shag. Both birds nest on cliffs and appear to hunt for fish in the same waters. Detailed studies have shown that, in fact, the two birds can coexist in the same area because they occupy different niches. Shags hunt in shallow water, eating fish such as eels and sprats, while common cormorants take fish from deeper water, feeding from shrimps and cod. The nesting sites of the two birds are also different. Common cormorants build their nests on cliff tops and broad ledges while shags nest on low boulders and narrow ledges.

Careful observations of five different species of warbler in North America, all of which seem to share the same niche, have produced similar results. They were studied in the 1950s by the ecological theorist, Robert MacArthur. He noted that a single spruce tree might be inhabited by all five species of warbler,

Unlike common cormorants, shags, on the right, prefer to nest on narrow ledges and eat mostly sand eels and sprats, which common cormorants avoid.

This pair of common cormorants has built a nest on a cliff. Fishermen used to shoot these birds, believing that they reduced fish stocks. In fact, common cormorants eat mainly shrimp and only a few commercial fish species.

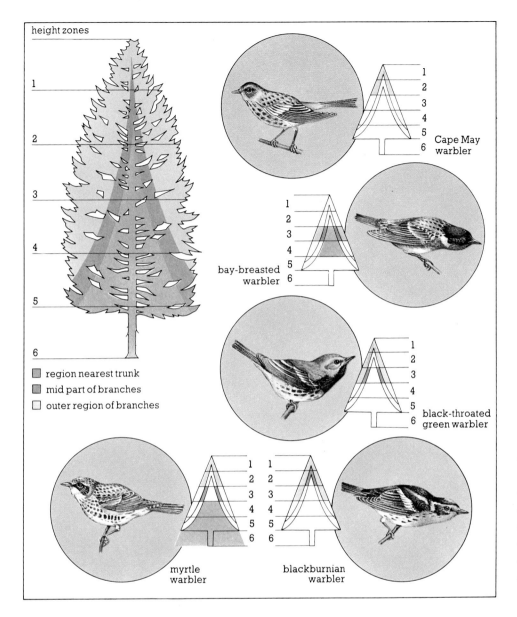

The Cape May warbler, the bay-breasted warbler, the black-throated green warbler, the blackburnian warbler and the myrtle warbler can all live in one kind of spruce tree in North America because they hunt mainly in different zones, indicated here by the coloured sections on the trees.

feeding on bud worms, a type of caterpillar that is found on spruces. Each species of warbler, however, fed mainly within its own territory.

Sometimes species have overlapping niches and share some of the same resources, but coexist because their requirements are not absolutely identical. Swallows, swifts, and martins are common summer migrant birds in most of North America and in Europe. They are all extremely agile fliers that eat flying insects and parachuting spiders, and they frequently fly in mixed flocks hunting the same prey. Although their niches do overlap they coexist because they have different nesting sites. Each type of bird also migrates to its seasonal sites at slightly different times of the year.

In different areas of the world there are similar relationships between plants (such as grass) and animals (such as the herbivores that eat grass). Since plant and animal distributions are rarely exactly the same worldwide the organisms that occupy similar niches in different areas will not be identical. They will, however, have similar functions, and are known as ecological equivalents. In the grassland habitats of Africa, for example, the major herbivores are species of antelope and zebra. In Australia the same niche is occupied by the large kangaroos, while in North America the equivalent used to be the bison.

Communities

The plant and animal populations that live in a specific area, or habitat, are known as a community. Some communities, such as those found in arctic tundra, contain only a few species while others, such as those in tropical rain forests, contain many thousands. Although communities may differ in the number of species that they contain, they are recognizable as separate units because their plant and animal populations can survive independently of those in adjacent communities.

In any community a few types of plants and animals are very common while many others are relatively rare. The most abundant species of plants or animals, is called dominant, and it gives the community its character. Where the dominant species is grass, for example, ecologists speak of a grassland community and where corals are prevalent they refer to a coral reef community.

Communities are not random assortments of plant and animal populations. They show a high degree of organization that is based on how energy is transferred from one species to another. In a small grassland

Found only along the eastern coast of Australia, koalas spend most of their lives in the tree-tops, eating the leaves and young bark of twelve species of eucalyptus trees. Koalas are mostly solitary animals.

Saguaro cacti, which store large amounts of water in their prickly stems, are one of the drought-adapted plants that can live in this desert community in Arizona, U.S.A.

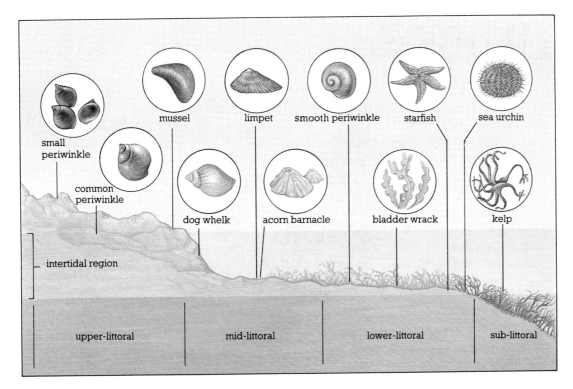

This diagram shows a community living on a rocky shore in the northern hemisphere. All the plants and animals are specially adapted to the saltwater waves and to periods without water, when the tide recedes. Winkles, mussels and barnacles attach themselves firmly to rocks and some plants, such as the bladder wrack, are accustomed to lack of water for a short time.

community a few species of grass produce most of the plant growth and provide most of the food – and thus the energy – for the dominant herbivores of the area, such as deer or rabbits. If the dominant grass species were removed, their place would be taken by other plants, which might be unsuitable for the dominant herbivores, and so cause them either to move away or to die. Other kinds of herbivores would then become dominant, thus altering the nature of the community. Dominant herbivores are themselves responsible for maintaining the relative abundance of plant species by the combined effects of grazing and trampling. If the herbivores were removed, the composition of the plant species would change. The numbers of carnivores that feed on the particular herbivore species also change in response to the availability of the herbivore and the plants on which the herbivores feed.

The less abundant plants and animals of a community are not as important as the dominant species for maintaining the special characteristics of a community. Removing any of these organisms would have little impact on the other members of the community, as the proportion of food that the less important plants and animals produce or consume is only a fraction of the total amount of energy in a community. For example, a few orchids may grow in a vast deciduous wood. If the orchids were removed it would have no effect on the ability of the community to thrive.

Some communities have distinct boundaries. This is most obvious where the physical conditions of the habitats that the communities occupy are different, such as the boundary between land and water. Usually divisions between communities are not very well defined. The boundaries between forest and scrub, for example, often contain species from both areas.

There are recognizable relationships between organisms within any community. Different plants often grow close to each other because they require similar physical conditions. Some species have very specific requirements for moisture, others for sunlight and others for soil minerals. In a forest the large trees form a canopy that shuts out most of the light from the areas beneath them. Some shrubs are adapted to the lack of light and grow in the shade of the larger trees.

Many species of animals are associated with a particular plant species. These special relationships are most frequently seen in insects. The larvae of some butterflies are confined to a few kinds of plants. In the United States Monarch butterflies, for example, feed only on a few species of milkweeds. Among other, larger, animals, the Australian koala eats only the leaves and bark of particular types of eucalyptus trees.

Succession

When an area of land is cleared of its vegetation it is soon colonized by plants and by animals associated with the plants. Over a period of years different types of plants will gradually replace each other. This process of sequential change is known as succession.

Succession takes place whenever a new area of land is available for colonization. This area could be a coastal sand dune, or rock recently bared by a retreating glacier, or it may be where existing vegetation has been disturbed – when a large forest tree falls down or a new road is constructed, for example, or when there is a forest fire. When a barren area of land, such as a sand dune, is colonized, the process of vegetational change is called primary succession and takes place over hundreds of years. The return of an area to its natural vegetation after disturbance is termed secondary succession. In both processes there are recognizable stages. These stages are determined by the quantities of water, nutrients and light available for plant growth. At the same time, the vegetation of the various stages may also affect the physical factors.

A good example of primary succession is the change that takes place on a newly formed coastal sand dune over a period of time. The new dune drains very quickly, holds little organic matter and its sand blows away easily. Most plants are unable to grow in such a harsh environment but a few, such as marram, or beach grass, can establish themselves at the edge of the dune and send out a network of rhizomes (underground stems with buds on them) under the sand, from which new shoots emerge. The large root system enables the grass to absorb enough water to protect the dune from erosion, thus stabilizing it. As the grass dies, decaying organic matter, or detritus, accumulates in the sand. The detritus contributes nutrients to the sand and helps the sand to hold in water.

This photograph shows a view toward South Harting from the hills of the South Downs in Sussex. The fields in the background, if left untended by people, would be colonized by tall grasses, annual plants and shrubs, as in the foreground. This kind of vegetation would eventually be succeeded by trees, such as the whitebeams, beeches and yew in the middle distance.

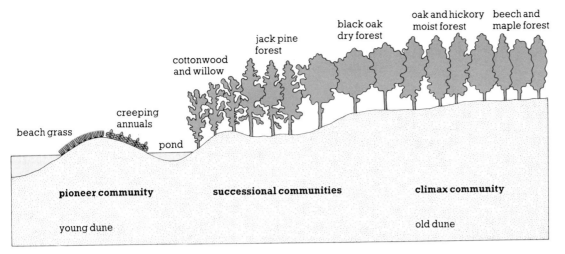

The process of succession taking place on the sand dunes of Lake Michigan is shown in this diagram. Some trees, such as jack pines, grow in areas affected by forest fires.

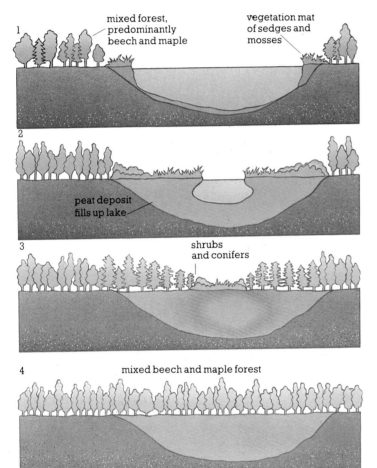

This diagram shows a succession from lake to bog to forest. At the edge of the lake a mat of vegetation forms. These sedges and mosses encroach on the lake, which gradually fills up, allowing shrubs and conifers, and, eventually, beeches and maples, to grow.

These changing conditions on the dune enable numerous annuals, those plants that live for only a year, to germinate and grow. Annuals, in turn, add more detritus when they die, increasing the water-holding capacity of the dune and also adding to its supply of nutrients. Gradually the conditions on the dune become suitable for the growth of perennials, those plants that live for long periods, such as woody shrubs and willow. The shade cast on the ground by the shrubs prevents the growth of annuals. The conditions that perennials create enable tree species to grow. Eventually woodland or forest may develop.

When an area of vegetation is disturbed, the soil itself is seldom removed, and so one of the first stages of primary succession – the creation of humus – does not occur. When an agricultural field is abandoned, for example, the first colonizers are annual weeds. Consequently, secondary succession is a much faster process than primary succession. It is estimated that it has taken about a thousand years for forests to develop on the sand dunes around Lake Michigan in North America whereas similar forest has grown up on agricultural land within only two hundred years in the same region.

Occasionally, areas of vegetation are disturbed so much that secondary succession cannot take place. In some places, such as southern France near the Mediterranean Sea, severe forest fires often burn away most of the humus – which holds plant nutrients – from the soil. Plants that require a great deal of nutrients are therefore unable to colonize the area and so the successional process resembles primary succession instead.

Climax Communities

Some communities of plants and animals, such as those of the arctic tundra and tropical rain forests, are remarkably stable. They will survive indefinitely provided there are no changes in climate or disturbances, such as clearance. Such communities are known as climatic climax communities. Other plant and animal associations are unstable. Some prairie land in the American Midwest, for example, would revert to forest if it were not periodically burned. These prairies, and other similar areas, are successional stages that can develop into climax communities if they are left undisturbed.

In many areas climatic communities do not develop because of a particular local soil. In the coastal region of northern California, for example, the soil has developed from sandstone and the climatic climax community of the area is giant redwood forest. Nearby, there is an area of the sandstone that has developed a hardpan, a layer that restricts root growth and the movement of water and dissolved minerals in the soil. Redwoods are unable to grow in the hardpan and the climax community there is pygmy coniferous woodland. Climax communities that are modified by local soil conditions, such as the pygmy conifers, are known as edaphic communities.

A local climate may also affect climax communities. In the northern hemisphere the Sun shines from the south, so south-facing slopes of land are usually warmer than the average regional climate while north-facing slopes are cooler. In the subarctic, between tundra and coniferous forests, the climatic climax is a woodland consisting of low-growing dwarf birches and willows. On north-facing slopes the local climate is too cold for these plants to grow and the climax resembles that of arctic tundra with grasses, sedges and mosses being the dominant plants. On warmer, south-facing slopes the climax community resembles more southern regions and the vegetation is dominated by coniferous trees.

True climatic climax communities are rare and many plant and animal groups are maintained, or have been

This giant redwood in California has a diameter of four metres at its base. Such trees, which form the climax community in the sandstone soils of this coastal region, often grow to over ninety-eight metres high.

The pygmy cypress trees on the opposite page are growing on the hardpan soils of the northwestern mountains of California, not far from the giant redwood shown on this page. The pygmy cypresses in the foreground form the climax community of this area. They are fully grown trees, but grow no taller than the height of an average person.

This North American beaver is carrying a stick to its lodge. During the long, cold winters beavers store young twigs in their lodges as a food supply. The beavers' habit of building lodges and dams in different places periodically creates changing communities, called cyclic climax communities.

modified, by human interference and grazing animals. The savanna of Africa was once forest but was cleared by early cultivators whose agricultural activities greatly affected the soil. These areas are now dry grasslands subject to periodic fires caused by lightning and burning by farmers. Much of the region now only supports types of grasses and shrubs that are tolerant of both drought and fire. Because of the modified soil, the frequent fires, and grazing by wild and domestic animals, the establishment of forest trees is now unlikely. Significantly altered communities, such as the savanna, are often thought of as a type of climax, and are called artificial climax communities.

Within climatic communities there are natural events that alter parts of them. In the cool northern forests of the world the climax community is dominated by birches, maples and firs. Beavers living in the rivers there construct dams and lodges (their homes) by felling trees with their strong and sharp incisor teeth. Both the male and female beavers help to build lodges, which vary in shape and size, and usually contain several 'rooms'. The dams, up to 300 metres long, create large spaces of open water. Trees in the waterlogged area quickly die, and a bog develops. In time the bog will revert to forest through succession. A family of beavers will abandon an area when they have exhausted its food supply and build a new dam and lodge somewhere else. Such a system of changing communities can be considered a cyclic climax, the cycle being controlled by the beaver.

The Ecosystem

The community of plants and animals that inhabit a defined area, together with their abiotic, or non-living, environment, is known as an ecosystem. Ecosystems are highly organized units, relatively self-sufficient in nutrients, and capable of independent existence from other ecosystems. All ecosystems, however, require an external force of energy – usually sunlight – in order to function.

In a pond ecosystem, for example, most of the chemical and physical elements necessary for life, such as nitrogen and carbon, are recycled within the pond and so retained in the ecosystem. The pond also depends on some outside components and forces, for its existence. Water that evaporates from a pond, for example, is replaced by rainfall or water draining off the land, called surface run-off.

The living parts of an ecosystem are linked by feeding relationships within it. Some plants and animals, however, may also be eaten by animals from other ecosystems, and some animals from one ecosystem eat organisms from others. In a pond, small fish and frogs are eaten by herons, which also feed in other aquatic ecosystems. Amphibians, those animals that can live either on land or in water, feed on flies and other insects from land ecosystems. When amphibians die and decompose in a pond they contribute energy to the pond from the food that they have eaten on land.

Although ecosystems may look different, they are all organized in a similar way. In a pond or a lake, for example, the non-living components are chiefly water and sediments. The atmosphere also plays a role by supplying emergent vegetation, such as reeds, with carbon dioxide. A limited exchange of oxygen and carbon dioxide also takes place between the water and the atmosphere. Most of the primary producers in a pond are microscopic green algae, which are suspended in the water. Other producers, such as lilies, are rooted in the sediments.

In a land ecosystem, such as a forest, the non-living components are the mineral soil and the atmosphere. The primary producers are green plants that are rooted in the soil although a few, such as lichens and some mosses, are attached to the branches and trunks of trees.

The animals that inhabit land and water ecosystems fall into several different categories, determined by the type of food they eat. Plant-eating animals, or herbivores, are called the primary consumers. Herbivores cover a wide variety of animals, including zooplankton and some types of insects in water, and insects and rodents on land. The secondary consumers are the meat-eaters, or carnivores. Carnivores include some insects, amphibians and many fish in water, and many kinds of animals (such as hawks and cats) on land.

Although land and water ecosystems differ in their components, the organization of both ecosystems is the same, as shown on the right, and both depend on the energy of sunlight to function.

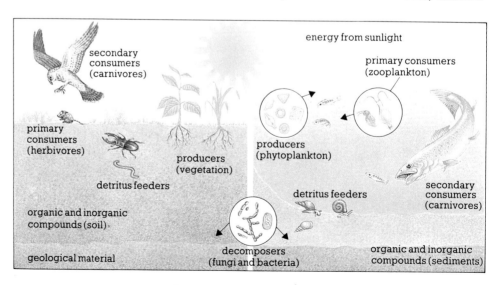

structure		function
inorganic (nonliving) substances, e.g. water (right), gases and minerals	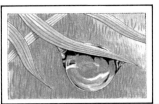	They are essential for life and are incorporated into the tissues of micro-organisms, plants and animals.
living producers green plants		They build complex substances, such as starches and proteins, from simple, inorganic substances, such as oxygen, carbon dioxide and nitrogen.
consumers (i) herbivores (primary consumers), e.g. deer (right), rabbits, snails, sardines		They feed directly on green plants. Herbivores provide food for other consumers, including parasites and scavengers.
(ii) carnivores (secondary consumers), e.g. weasel (right), lions, eagles, sharks		They feed directly on the bodies of herbivores (and sometimes other carnivores) and thus feed indirectly on green plants.
(iii) omnivores (primary of secondary consumers), e.g. fox (right), humans, rats, herring gulls		They are opportunists who will exploit any available food source, plant or animal.
(iv) parasites, e.g. tick (right), fleas, tapeworms, gall wasps, dodder		They feed on, or within, the living bodies of animals.
(iv) scavengers, e.g. vulture (right), hyenas		They are also opportunists, and eat carrion (dead animals).
decomposers e.g. fungi, such as orange peel fungus (right), and bacteria	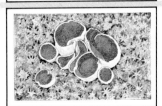	They feed directly on the bodies of herbivores (and sometimes other carnivores) and thus feed indirectly on green plants.

This diagram shows the different parts of any ecosystem, and indicates how they work, and their relationships to each other.

A water beetle, shown here taking in air, spends most of its time in ponds where it feeds on tadpoles, insects and tiny fish. It is an example of an animal that moves freely from one kind of ecosystem to another; although it lives most of the time in water, it has to move out of the water in order to breathe. The air enters its body through openings in its abdomen, which it projects upwards into the air, as in the photograph.

Energy in Ecosystems

Unlike nutrients, which are cycled between living and non-living forms, energy flows through the ecosystem, passing from sunlight through plants to animals. Animals may obtain chemical energy from plants either directly by eating plants or indirectly by eating other animals. It is eventually lost from the ecosystem in the form of heat.

Energy passes between the food producers and consuming animals in a series of ordered steps, called trophic levels. There are producers in every ecosystem. These are green plants that obtain their energy from the Sun and convert this energy into forms that other organisms can use, namely sugar and starch. Plants make up the first trophic level. Plants are eaten by primary consumers, the herbivores, which form the second trophic level. The third trophic level is composed of secondary consumers that eat the herbivores. Consumers at this trophic level are usually carnivores and parasites, which are animals or plants that feed on living organisms. Other animals that eat secondary consumers are known as tertiary consumers and belong to the fourth trophic level. Many animals belong to different trophic levels. Bears eat berries and fish and are therefore both primary and higher consumers.

Energy is lost from the ecosystem at each trophic level. Every organism burns up energy, producing heat when it moves and respires. This means that only a small proportion of the energy consumed is used for growth and reproduction. The amount of energy available to any plant or animal is therefore less than that available to its prey. In African grasslands, for example, there are fewer lions and hyenas than gazelles and zebras because more energy is available to the herbivores (in the form of the grass that they eat) than to the carnivores.

The trophic levels of a community make up a food chain, depicting who eats what. Plants and animals rarely depend on just one kind of food and are seldom eaten by only one predator. Weasels, for example, may consume both herbivorous and insectivorous mice and shrews, and are themselves eaten by large owls and other birds of prey. It is rare to find a simple food chain in nature and so ecologists usually see the transfer of energy between organisms as a series of linked food chains, or a food web.

Energy is not always transferred from producing organisms to consuming organisms. Some plants consume animals. The Venus's flytrap, pitcher plant, and sundew are all examples of insectivorous plants. Although they use the energy of sunlight for photosynthesis and are primary producers, they also digest small insects and are therefore consumers as well.

In most land ecosystems the major pathway of energy is not, as might be thought, through the grazing food chain but through the detritus food chain. Detritus feeders are also found in aquatic ecosystems. These organisms consume the remains of dead plants and animals and their waste products. Most detritus feeders, such as worms and bacteria, are small, but they also include some large animals, such as vultures and crabs. The role of detritus feeders is to break down large remains into smaller pieces that bacteria and fungi decomposers can use as a source of food and energy. The final products of these decomposers are soluble nutrients, which can be absorbed by plants, and waste energy, which is lost as heat.

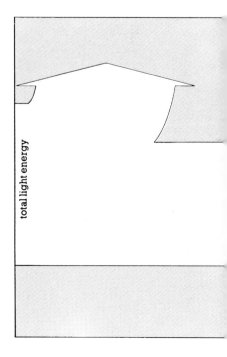

The diagram shows energy flowing through the trophic levels. At each level a large percentage of energy is lost (from forty to ninety per cent) through respiration, decay or excreta. Net production of energy is therefore very small, and decreases as it moves up the trophic levels.

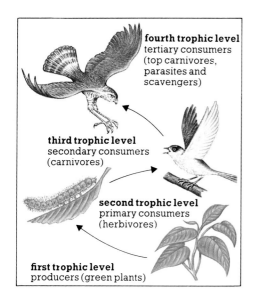

The simple food chain above shows examples of organisms from the various trophic levels.

The diagram on the right illustrates a food web in a pond and shows the feeding relationships between the different plants, insects and fish living in the water.

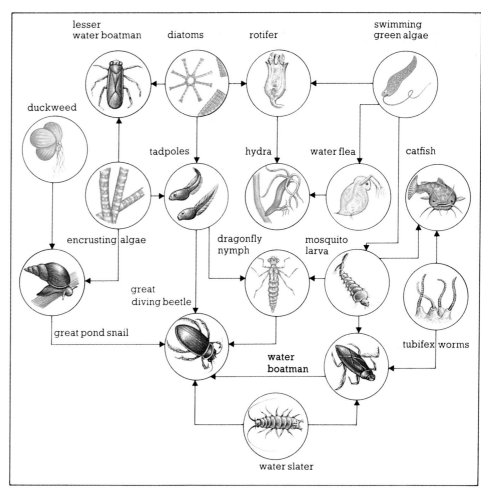

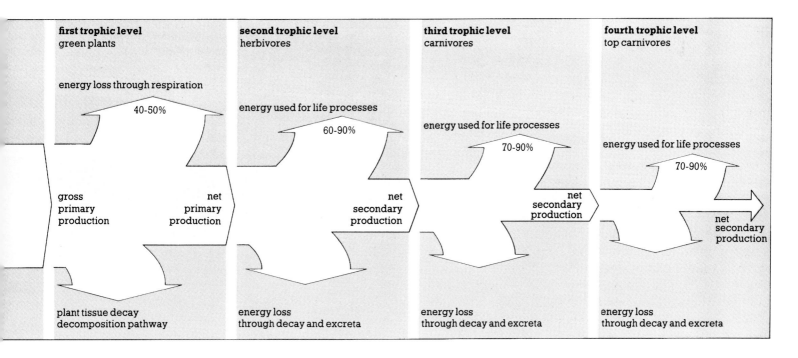

31

Ecological Pyramids

The number of animals that can live in a particular area depends on the amount of food that is available for them there, and also on the methods they have to use to get their food. Predators must be large enough to capture and kill their prey. Thus foxes can easily kill rabbits but are not able to kill larger herbivores such as deer, which are too large for them to catch. Some predators, including hyenas and wolves, hunt in packs. The combined strength of the pack enables the predators to capture prey larger than themselves, such as zebra and wildebeests. Unlike carnivores, herbivores do not have to be very large or strong to obtain their food; so their size is seldom related to their food. Aphids, for example, are tiny but elephants are huge.

As energy decreases along a food chain there are usually fewer individuals at each succeeding level in the chain. In a grassland community the number of plants is much greater than the number of herbivores. The plant eaters are in turn more numerous than their predators, such as birds and cats. These decreasing numbers can be shown by plotting a graph in the shape of a pyramid. In certain communities this pyramid of numbers can be inverted, with more individuals at a higher level in the food chain than at a lower level. In a forest, for example, there may be fewer primary producers, such as trees, than primary consumers, such as insects.

If the amounts of energy available to the succeeding

This pyramid graph reveals how the number of individuals in a food chain decreases at each succeeding trophic level, while the size of the individual tends to increase. Examples of the individuals in such a chain – green plants, beetles, shrews and an owl – are shown on the right.

This African elephant, an example of a very large herbivore, uses its mobile trunk to convey food and water into its mouth. The tusks are upper incisor teeth that continue to grow throughout the elephant's life. Elephants, because of their size, have few enemies except humans, and are now a protected species.

groups of consumers are represented as a block, a pyramid of productivity, rather than of numbers, is produced. These blocks will always be pyramidal in shape but the proportion in each one will vary between communities because some organisms are more efficient at turning food into growth than others. Chickens and other fowl are, for example, far more efficient than cattle at turning food into flesh.

An alternative method of showing how the organisms of food chains are related to one another is to plot a pyramid of biomass. Biomass is, literally, the living weight of the organisms at different levels in the food chain. This living weight drops step by step just as the energy available to organisms decreases in food chains according to the number of animals that the energy has passed through.

A pyramid of biomass will resemble a pyramid of productivity in most land-based communities. In a grassland community, for example, the combined weight of plant material will outweigh the combined weight of all the herbivores. The weight of all the herbivores will, in turn, be heavier than that of all the carnivores. Although individuals that are high in the food chain may outweigh those lower down in the chain, biomass pyramids are concerned with the combined weights of individuals. A woodpecker is heavier than its prey, a bark-living insect, but the biomass of woodpeckers is less than the combined weight of all of its prey because woodpeckers are thinly dispersed throughout a forest while its prey exists in large numbers. Similarly, an owl is heavier than a shrew, but the large number of shrews in a habitat will outweigh the few owls that may live there.

This buzzard (Buteo buteo) *is a top carnivore and has swooped down from high up in the sky to attack its prey, a rabbit, which is a herbivore and therefore a primary consumer.*

How Nature Controls Populations

If a culture of small organisms is kept in the laboratory with a limited amount of food it will quickly grow in size until the food is exhausted. Without more food the culture will die. If the food is replaced periodically, one of two things will happen to the size of the population. The number of organisms may increase and be maintained at a constant level, with minor fluctuations in numbers. Alternatively, the culture may undergo a series of population explosions, the size of the population swinging from very few individuals to a great many. Such patterns are often observed in nature.

Population explosion cycles happen most frequently in simple ecosystems where few species live. In the Eurasian and Canadian tundra lemmings are well known for their regular three to four year cycles of abundance. When the population reaches very high levels the lemmings become overcrowded and react by mass emigration to find new food, and this often leads to their death. The population then collapses, only to increase again later and then repeat the cycle. The cycles of the lemming population correspond to cycles of its main predators: arctic foxes and snowy owls. Similar cycles have been observed in arctic hares and lynx.

Some plant and animal populations are regulated by mechanisms that are most effective when populations are large and have little effect when populations are small. They are known as density-dependent mechanisms. One such mechanism is territorial behaviour, which is found in many different kinds of animals. Many species of song-birds establish territories during the breeding season. One male robin, for example, will establish itself in a territory where enough food is available to raise a brood of young. The male attracts a mate to its territory and deters others from stealing both the territory and its mate by singing. Occasionally encroaching males are deterred by mock fights. The size of territory varies from year to year, being larger when there is little food and smaller when it is plentiful.

These lemmings are migrating in Sweden. They often drown when they try to cross rivers, as in the photograph, or reach the open sea.

As robins will only breed if they are in a territory, the population size is regulated in a manner that corresponds to the availability of food. This ensures that robin populations do not grow to a size that is larger than their food resources can support.

Other density-dependent mechanisms involve the control of population size by external organisms. The spruce sawfly was introduced by accident to Canada

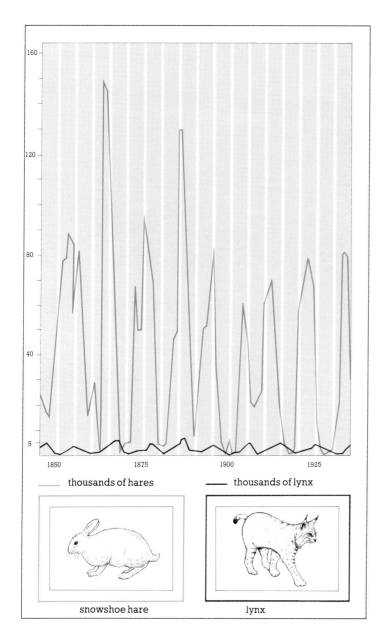

thousands of hares — thousands of lynx

snowshoe hare · lynx

This graph shows how the populations of arctic lynx and their prey, snowshoe hares, fluctuated between 1845 and 1935. The graph reveals that the number of lynx rose whenever hares were numerous, and fell whenever hares were scarce, in a constantly repeating cycle. The lynx decrease in about 1850, for example, followed the hare decrease in the late 1840s. The Hudson's Bay Company in Canada kept the records on which this graph is based.

In Kenya, a lioness attacks a wildebeest that has strayed away from the main herd. Lionesses, rather than lions, do most of the killing on the savanna, and often form teams in order to stalk their prey.

from Europe about 1930. It rapidly invaded vast areas of eastern Canada and the United States and by 1937 had defoliated between half and two-thirds of the white spruce in an area of 7767 square kilometres. Entomologists, who study insects, introduced several parasites in an attempt to control the sawfly, but with little effect. Eventually a virus was found to be the most effective control. When the population increases the virus spreads rapidly, causing a reduction in numbers. When the sawfly is low in numbers the virus has little effect, as it cannot readily spread through the population. The sawfly population is now regulated by this virus in a density-dependent manner.

Predators are not always an effective way of regulating populations. In the African grasslands and savanna regions there are large herds of grazing animals including zebras and gazelles. Their predators, such as hyenas and lions, take mainly sick and lame animals that cannot keep up with the herds. These animals would die even if they were not taken by predators.

Not all population regulation is caused by living factors. Some species, particularly those that are small or are located in unstable environments, may be regulated by natural agents, such as fires and floods.

Competition and Coexistence

During the 1930s the Russian biologist G. F. Gause conducted some simple experiments with microscopic animals called *Paramecium* to determine whether two animals that consume the same food could live in the same habitat. First he kept cultures of different species of *Paramecium*, with the same food, in separate test tubes. Gause found that they could live almost indefinitely provided the food was replaced periodically. When he mixed two different species of *Paramecium*, and maintained the mixed culture under the original conditions, he found that only one species survived. However many times the experiment was repeated it was always the same species that survived. Gause also discovered in another experiment that two particular species could survive together because they lived in different parts of the test tube. These two species coexisted because they had partitioned their living space. Other mixed cultures competed for the same living space and therefore did not coexist. The principle that no two species can occupy the same niche – that is, use the same food and living space – is universal. It is often called Gause's principle of competitive exclusion.

Similar patterns to Gause's experiments can be seen in nature. Competition happens when two or more species require the same resource, such as food or a space to live, but there is not enough of the resource for more than one species to use. On the rocky coast of Scotland, for example, there are two related barnacle species called *Chthamalus* and *Balanus* that live on the rocks between low and high tide marks. Barnacle larvae (the young stage of barnacles) are evenly distributed between the tide marks. The adult barnacles are not evenly distributed, however. *Balanus* lives in the

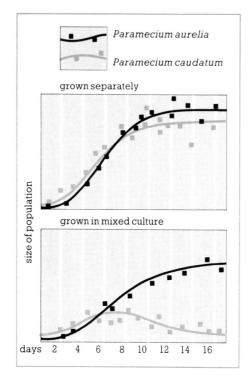

Gause's graph, right, shows the results of his experiments with cultures of two different species of Paramecium. *The top graph shows how both species increase at almost the same rate when they are separated. When they are grown together, however,* Paramecium aurelia *excludes* Paramecium caudatum, *causing the numbers of* Paramecium caudatum *to decline, as shown in the lower graph.*

Far right: large Balanus *and small* Chthalamus *barnacles are seen here on an exposed rock at low tide. Barnacles are related to crayfish and lobsters despite the fact that they attach themselves to one place.*

lower half of the habitat and *Chthamalus* in the upper half. There is always a sharp boundary between the two species. *Chthamalus* can grow over the whole intertidal zone but *Balanus* cannot grow in the upper, drier, zone. The two species compete for rock space in this lower zone but *Balanus* always wins because it grows faster than *Chthamalus* and, in doing so, physically prises the young *Chthamalus* off the rock.

When animals or plants are introduced into areas where they were not living before they often compete with the native plants and animals for the same resources. The rabbit was first introduced into Australia in 1859 and its population exploded in only a few years. The rabbit is a grass-feeding herbivore and the increase in its population size caused a decline in the quality of the grasses that the native herbivores, such as kangaroos and wallabies, also ate. Until effective rabbit control measures, such as fencing, were introduced in the 1930s the competition between the native herbivores and rabbits was so severe that in some areas the original herbivores were almost driven to extinction.

Not all species that use the same resources are in competition. In African grasslands there are mixed herds of zebras and wildebeests that feed on various grasses. These zebras and wildebeests are followed by herds of Thompson's gazelles. In the 1960s, examination of the stomach contents of these animals proved that each of them ate different types of vegetation. Zebras snip off long dry grass stems with their very sharp incisor teeth and wildebeests tear off grass sideshoots with their tongues. Together these two species act cooperatively as they each expose, by their method of feeding, the parts of grass that the other species eats. The Thompson's gazelle eats small plants exposed by the feeding habits of the other two species.

In this photograph zebras and wildebeests are feeding on the grasslands at Ngorongoro in Tanzania. There are several species of zebras, each having slightly different markings. It is not known why zebras are striped – their markings are probably not for camouflage, as the grasslands where they feed are open and have few shrubs and trees.

Close Relationships in Nature

The terms 'struggle for existence' and 'survival of the fittest' evoke images of fierce animals, such as lions and tigers, devouring helpless prey. Not all relationships between different species are like this. There are many examples of organisms interacting in different ways, often to their mutual advantage.

Some plants and animals are totally dependent on one other species for all their requirements. An organism that provides another with all its needs is called a host. When the host gains no advantage or is harmed, the relationship is known as parasitism. Parasites are usually smaller than their hosts and infrequently cause their death because this would also cause the death of the parasite. Stable relationships between parasites and their hosts have developed over long periods and if a new parasite is introduced to a host it may kill the host.

Many plants and animals are beneficial to each other. Such relationships are described as mutualistic. The mohave yucca plant, found in North and Central American deserts, is pollinated by one insect, the yucca moth. The larvae of this moth are also totally dependent on the plant for food. Female yucca moths enter yucca flowers and deposit between one and five eggs on the ovary of the plant. The growing larvae eat the developing yucca seeds, but only about a third of them. The plant is able to withstand this level of seed predation. When the female moth leaves the flower it collects pollen, which it transfers to the next flower that it visits. Although the moth is a predator of the plant it is also its sole pollinator and so the relationship benefits both species.

Not all mutualistic relationships are as close as that of the yucca plant and the yucca moth. Many organisms that are capable of an independent existence will do better if they are associated with others. In the Bahaman waters there are numerous shrimps, including the Pederson shrimp, which is well known for its activity of 'grooming' fish. Many fish search for these

This female yucca moth is pollinating a yucca flower in the Mexican desert.

Cladonia *lichens, such as those below, are formed by the close relationships of algae and fungi. They are very sensitive to pollution in the atmosphere and are seldom found where there are high concentrations of sulphur dioxide in industrial areas.*

shrimps and when they locate the shrimp they allow it to pull off parasites and areas of injured or dead skin, which would otherwise develop fungal infections. Fish would be capable of independent existence without the shrimp but do better if they are associated with it.

Many animals have relationships with plants because the animals resemble a particular plant and use this similarity to hide either from predators or their prey. Some wolf spiders closely resemble flower parts, mainly the petals. The spiders sit in flowers undetected by pollinating insects. Insects that land on the flower do not see the spider and are quickly eaten by it. Many insects, on the other hand, also avoid detection by their predators because they resemble plants themselves. The Comma butterfly looks like a dead leaf on its underside and hibernates or rests among dead foliage. Any predator will overlook it because the butterfly merges into its background.

Some animals gain protection from predators because they resemble a distasteful or poisonous species. In Trinidad many Queen butterflies have similar markings to Monarch butterflies and so are protected from bird predators. Although the Queen butterflies are edible, birds associate them with the unpleasant Monarch because they look the same.

Some relationships are so close that two species may become fused together. Lichens, which form greyish-green crusts on trees and stones, are often thought of as a group of plants. Lichens are, in fact, intimate associations between fungi and algae, and it is often impossible to separate them into their fungal and algal parts. The algae convert nitrogen into nitrates and produce sugars that the fungi can use, while the fungi supply the algae with water and food. Lichens obtain water partly from rain, partly direct from the air, and partly from dews and fogs.

This diagram shows the life cycle of the parasitic liver fluke and its series of hosts. Zebra or heather snails (1) are the first hosts. These snails become infected by eating fluke eggs within the faeces of sheep, roedeer and wild rabbits. The parasite reproduces within the snail and fluke larvae develop (2), to be released by the snail in balls of mucus (3). Ants, the second hosts, eat the mucus balls. The infected ants, with the developing fluke inside (4 and 5) attach themselves to the tips of plants by biting them, and are then eaten by the final hosts, sheep, roedeer and wild rabbits, as they graze. The fluke larvae grow to adulthood in the livers of the final hosts (6, 7 and 8).

The Diversity of the Earth

The world's vegetation can be divided into various zones that correspond to the different areas of climate. These zones are known as biomes. There are eleven major world biomes including deserts, tundra and temperate forests. Each zone is unique for the plants and animals that it contains.

Sunlight, temperature, water vapour and precipitation are responsible for the climate of a particular area. The Sun, for example, is angled directly above the tropics, and these are the hottest parts of the world. Rainfall is heavy in the tropics and the atmosphere is very humid because warm air can hold a large amount of water vapour. These conditions support rain forest vegetation. By contrast, the Sun is at an oblique angle at the Poles and these areas are correspondingly cold, with little or no vegetation.

The Sun's energy also causes major atmospheric air movements. Warm air rises at the tropics and cools as it moves upwards into the atmosphere. The water vapour in the air also cools, condenses, and then falls as rain. The cool air is cycled over to the subtropics where it moves downwards, warms up, and absorbs water vapour. The warmed air, containing water vapour, is then cycled back to the tropics. The world's major deserts in both the northern and southern hemispheres are found in the regions where cool air descends between the latitudes of 15° and 30°, and takes away the water vapour.

The seasons exist because the earth's axis is tilted towards the Sun. In the north at 60° latitude, where coniferous forests grow, the coldest and warmest months differ by as much as 26°C. In the tropics the difference is as little as 2°C. Rainfall also varies according to the seasonal position of the Sun. In India, for example, the heaviest rainfall occurs between May and October.

Large areas of land and mountains also affect climate. The southern hemisphere, for example, receives more rainfall than the northern hemisphere because it has a greater area of sea, and more water evaporates from the sea than from land. When air approaches mountains, it moves upwards. It then cools and the water vapour in the air is condensed, causing rain or snow to fall on that side of the mountain. Some of the air crosses over the mountain and moves downwards on the opposite side. This air warms up as it approaches the ground and absorbs – and holds – water vapour. This causes a dry area to be formed, known as a rain-shadow. Both the Gobi desert in Asia and the Great Basin in the western United States are examples of rain-shadows.

High mountains are cool on their upper slopes. On the higher slopes, the plants and animals begin to resemble those found in the cold regions of the world. The graduated changes in communities caused by varying physical conditions (such as temperature and

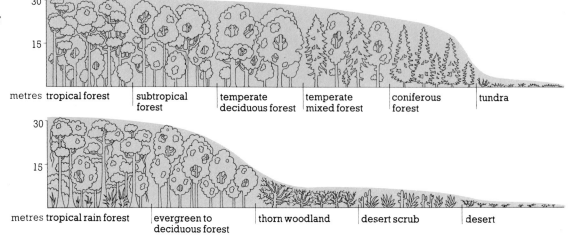

The diagrams on the right show two examples of ecoclines. The one at the top shows a gradient of decreasing temperature in the northern hemisphere. The lower diagram shows a gradient of decreasing moisture in South America.

humidity) are called ecoclines. From west to east central Africa, for example, there is a transition from rain forest to scrub – creating an ecocline along a gradient of decreasing moisture.

There are fewer variations in plants and animals in the oceans than on land, though different plants and animals do exist according to the depth of the water and the speed of the ocean currents. There is little seasonal variation in temperature within the oceans, but different areas of water may vary in temperature because of the movement of warm and cold currents.

Plants and animals are specially adapted for living in their biome. Many trees are deciduous, for example, and shed leaves when there is no water in the soil.

This map of the Earth shows the major biomes, or zones of different vegetation. Maps can only show such zones approximately since belts of vegetation, which are left undisturbed, seldom have definite boundaries but graduate from one type to another.

Some animals, such as the Arctic tern, are adapted to seasonal climates by migration. The tern raises its young in the Arctic during the short summer and migrates to the Antarctic oceans, where food is plentiful, during the winter months.

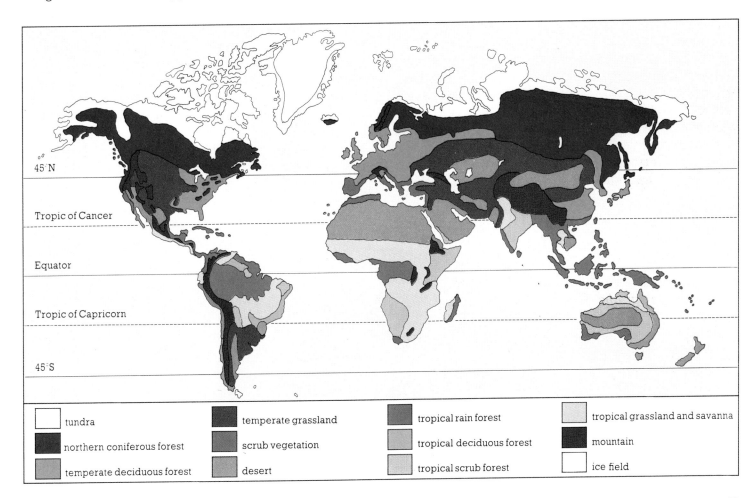

41

Freshwater Ecosystems

Fresh water is water that contains little salt. There are two major types of freshwater habitat – still water, such as ponds and lakes, and running water, such as streams and rivers. Conditions in both habitats influence the amount and the kinds of plants and animals that live in them. Still water, for example, often has limited supplies of dissolved oxygen and carbon dioxide, whereas oxygen and carbon dioxide are plentiful in running water. Similarly, all running water has a current of varying speed but still water has none. The transparency and the temperature of the water affect plants and animals in both habitats.

The most important plants in still water, those that are responsible for most of the primary production, are floating algae. These algae are found mainly on the surface of the water. Rooted plants, such as arrow grass, grow in places where enough light reaches the floor of the lake or pond.

In running water, most of the primary production is made by algae that are attached to surfaces, and encrust the rocks on the banks and floors of streams and rivers. Floating plants would be quickly washed away in running water. Plants are not essential for animal life in running water because of the currents, which keep organisms constantly on the move. There may be no plants at all in some rivers that have a great deal of sediment. The animals of such rivers exist entirely on the organisms and nutrients carried by the current from other water habitats, or which have fallen into the water from the land.

In shallow ponds and lakes the temperature remains the same at all depths, and gases and nutrients circulate freely. In deeper lakes the temperature and the concentration of nutrients and gases in the water vary with depth. Water has a different density at different temperatures. At 3.9°C water is cool and at its most dense, and warmer water floats on it.

In temperate and subarctic regions the heat of the Sun varies with the season. In the autumn, when the Sun's power is weak, the water on the surface cools and sinks to the bottom. In spring the surface water warms up, often from temperatures that are below freezing. The water at the top sinks again because the temperature of most of the water is at 3.9°C. The entire body of water is therefore cycled twice a year, distributing nutrients from the bottom layers. In tropical regions there is relatively little nutrient cycling in lakes because the surface layers vary little in temperature. Many tropical lakes are unproductive for this reason, and have few plants in them.

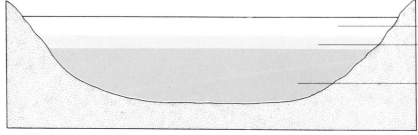

In a temperate lake in summer the upper layer of water is warm. The temperature rapidly drops in the second layer as the depth increases. The lowest layer of the lake contains cool water.

- surface lake (epilimnion) 21.1°C
- rapid temperature change (thermocline) 21.1-3.9°C
- lower lake (hypolimnion) 3.9--1.1°C

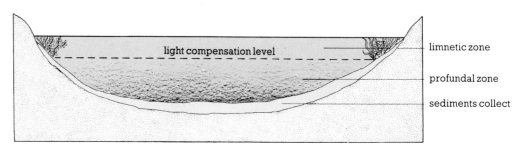

The section on the right shows the levels of light in a lake or a large pond. Photosynthesis can take place near the top, where plenty of light can penetrate, but not in the lowest, darkest zone.

- limnetic zone
- profundal zone
- sediments collect

Lakes that are low in nutrients, called oligotrophic lakes, contain few plants and much oxygen. Such lakes, including those of Canada and Sweden, are noted for trout and other game fish that like rich supplies of oxygen. Lakes that are rich in nutrients, called eutrophic lakes, have many plants but are low in oxygen because they contain large amounts of bacteria, which decompose plant remains, and take in a great deal of the available oxygen. Lakes such as these are often shallow or have been created from oligotrophic lakes by the addition of fertilizers, washed from agricultural land. Some animals, including diving beetles, can survive eutrophic conditions because they breathe air but others, such as fish, cannot live there because they rely on dissolved oxygen in the water.

All freshwater animals are specially adapted to their habitat. Animals that live in streams and rivers, such as stonefly larvae, are often flattened or streamlined, or live in areas protected from the current. The water-strider insect is adapted to living on the surface of water by skimming over it. This insect eats land-based animals, such as flies, that fall into water and are trapped by surface tension.

The South American lungfish has both gills and lungs. During periods of drought or when oxygen is low in the water, the fish can survive by breathing air with its lung. It also builds a mud cocoon in which it can live until water returns.

This stonefly insect larva is specially adapted to living in freshwater. The larva has long legs with two claws on each foot so that it can hold on to stones in fast moving streams; its flat body offers little resistance to water currents.

Marine Ecosystems

The oceans and seas cover seventy per cent of the Earth's surface. They are all connected, forming the largest, integrated ecosystem on the Earth. Most of the oceans are deep, the average being 3962 metres and, in some areas, much deeper. One oceanic trench in the Pacific goes down to about 10 972 metres deep. Life extends from the surface of the ocean to the floor.

Most life is found near the coasts where the waters are shallow, and nutrients and salts carried by rivers enter the sea. The nutrient content of these continental shelves, the shallow seas bordering shorelines, is also increased at the edges by the upward movement of nutrients from the depths of the oceans.

Ocean water is circulated all the time by warm and cold currents, and all oceanic regions and depths contain a large quantity of dissolved oxygen. The major zones of marine life occupy different depths that have varying temperatures, salt concentration, light levels and water pressure. Very few living things range over all the oceans, though it is possible that the blue whale may do so. Within the oceans there are some barriers to the free movement of organisms. Some can only live in warm currents. The Atlantic eel is found in the warm North Atlantic drift. Others are restricted to cold currents with ample nutrients, such as the Peruvian anchovy in the Humboldt current.

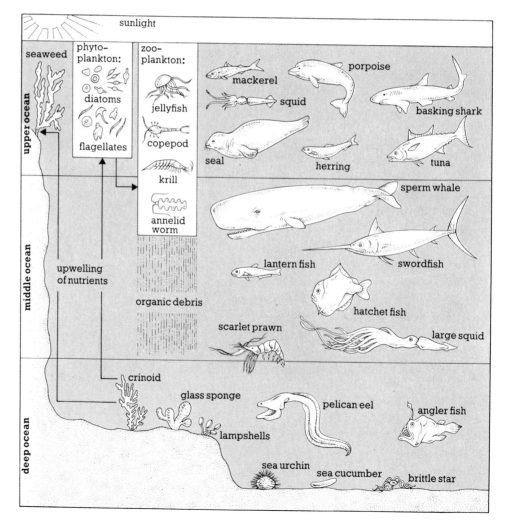

This diagram shows the basic food chain of the seas and oceans. Phytoplankton near the surface of the water are eaten by zooplankton, which are, in turn, eaten by fish. Some fish, such as sharks, prey upon others. Detritus sinks to the ocean floor where it is broken down by the decomposers. The depths of the oceans are also populated by specially adapted fish and other organisms. The nutrients released by the decomposers are carried upwards by circulating water, and are reabsorbed by the phytoplankton.

Salt concentration is extremely important to animals in the sea. If a solution of water and salt (as in seawater) is separated by a membrane, such as a fish's skin, the water will tend to move from the weaker solution to the more concentrated one. Salt will move more slowly in the opposite direction. This process is called osmosis. All sea animals have to balance the concentration of salt in their bodies to prevent the loss of water from their bodies by osmosis.

Many sea creatures, such as the jellyfish, have a salt concentration that is the same as seawater and do not suffer from the problems of osmosis. Fish, such as sharks and rays, secrete urea, the waste products of the food that they eat, into their bloodstream, raising their salt concentration until it equals that of seawater. Other fish, sea mammals and sea birds secrete salt from their specially adapted glands, balancing the movement of salt from the sea into their bodies. Fish and birds, such as terns, also drink seawater to replace the water lost in urine and from their body surface by osmosis.

The commonest marine plants, called phytoplankton, are microscopic diatoms, which float, and flagellates, which swim. Diatoms are more often found in northern waters while flagellates, which are often reddish in colour, dominate warmer waters. Some red-coloured flagellates are well known for producing red tides on the coasts of North America. They also produce a substance that is poisonous to fish and mammals. The commonest plants closer to the coast are multicellular algae, commonly known as seaweeds or kelps. They are usually attached to rocks.

The marine primary consumers are tiny animals known as zooplankton. These animals are minute jellyfish, small shrimps such as Antarctic krill, and the young of many fish, shellfish and worms. The primary consumers are eaten by small and medium-sized fish, and some mammals such as baleen whales. There are fewer marine consumers higher up in the food chains, and these include animals such as toothed sharks, killer whales, seals and tuna.

In the depths of the oceans there is little light and no plants. Most animals there are detritus feeders, but some are carnivores. Many deep-sea fish have peculiar adaptations for finding prey and mates. The anglerfish, for example, has a luminous lantern over its mouth to attract prey, and the pelican eel has jaws capable of swallowing very large fish.

Phytoplankton are most abundant near coastlines and cool currents where there are plenty of nutrients. Algae can be clearly seen in the photo below, taken near the Great Barrier Reef in Australia.

This photograph shows a dense group of tube worms, some up to three metres long, on the ocean floor. Among the tube worms there are also mussels and a crab. These creatures are part of a food chain that depends on sulphur produced by the chemical reactions of the Earth's crust, rather than sunlight, for their energy.

Life in Estuaries

An estuary is a stretch of coastal water that is partly enclosed by land, but connected to the sea. Estuaries also receive fresh water and nutrients from the land, either from rivers or from water, such as rainfall, draining off the ground. Coastal bays, tidal marshes, fjords (narrow strips of water between high cliffs), some deltas (triangular-shaped areas of land at the mouths of rivers), and lagoons are all examples of estuaries.

The conditions in an estuary are harsh. Plants and animals have to survive extremes of temperature, salt concentration, flooding, drying out, high levels of sediment and strong water currents. Estuaries are , however, extremely rich in nutrients because they trap those that come in from the sea as well as those from the land. The sea also removes waste products.

Estuaries contain three major kinds of plants. These are phytoplankton, algae that live in the sediments, and large plants such as seaweeds and grasses. All three types of plants are found in different parts of estuaries. Phytoplankton grow in channels that are permanently below the low tide level; sedimentary algae grow anywhere where there is sufficient light for photosynthesis and enough water to prevent drying out. Grasses, and similar plants, are found in the drier parts of estuaries, while some seaweeds grow in areas where there are at least two tidal floods a day.

Many estuaries can be divided into three zones. The communities that exist in these different zones can withstand varying water levels. Plants in the lowest zone, by the permanent channel of water, trap sediments around their roots and stems, and build up the height of the mud and silt. There is less tidal flooding higher up the banks. The plants of the wetter areas are replaced by others that grow in drier conditions. In a temperate salt marsh, for example, the pioneer plants, which trap sediments, are glassworts. Glassworts are replaced by plants such as sea-aster and these are, in turn, replaced by others such as sea-lavender, which grows in dry places. This process of succession is often interrupted by storms and floods that wash away the higher sediments, so causing a new succession to begin.

In order to combat the daily changes in salt concentration many small animals have shells that can be closed. Others, including some types of worms, have skin that restricts the passage of water in and out. Plants often have thick cuticles, and cord grasses secrete salts from their leaves. Many animals avoid drying out by burrowing and some, such as shrimps, follow the tides across the estuary.

Tides can wash the larval forms of animals out to sea, so many estuary creatures have unusual methods of reproduction. Young estuarine sandworms can swim at birth and then burrow into the mud. Shrimps attach their eggs to their bodies. When the eggs hatch the larvae are already able to swim.

A sandworm, on the right, lives in a burrow in mud or sand. Sandworms draw water through their burrows by moving their bodies with a wave-like motion that also pushes out waste.

Mussels, far right, are filter-feeding molluscs with hard, dark grey shells that attach themselves to rocks in estuaries. They are preyed upon by dog whelks, another hard-shelled animal.

Estuaries are important feeding grounds for many birds, fish and shellfish because of the abundant food in them. Eels often remain in estuaries in order to feed before they migrate from the sea to the rivers where they breed. Estuaries also provide food for man, including such edible plants as samphire, and shellfish, such as oysters and shrimps.

Knots are shore birds that feed on molluscs, crustaceans and worms, which they find in the mud halfway down the tide levels of estuaries. Knots migrate from Canada and spend the winter in Britain.

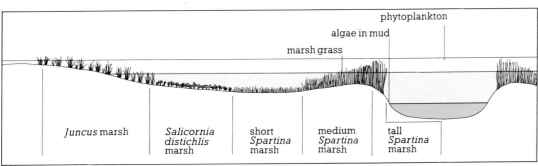

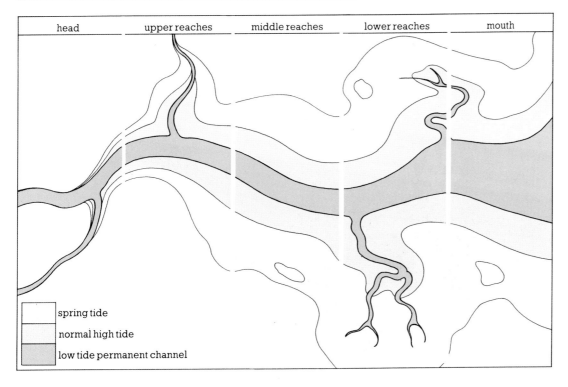

The top diagram shows a section through an estuary with examples of the plants unique to each succeeding zone. The bottom diagram shows a typical estuary viewed from above, and shows the varying levels of water.

Coral Reefs

Coral reefs, which are rich in animals and plants, are situated in clear, shallow, tropical waters – in the Pacific, the West Indies and Indonesia, for example. Reefs have been likened to an oasis in a desert. Nowhere else in the sea is there such a kaleidoscope of living things.

The main structure of a reef is formed by the remains of generations of lime-secreting organisms, especially the corals known as scleractinians. Calcium-secreting algae also play an important part in building reefs. Corals themselves are animals, similar in their basic structure to sea anemones, but differing from them because corals have a calcium carbonate skeleton.

The photograph, below left, shows a healthy coral reef in the Red Sea contrasted to a reef, below right, damaged by sediments from the Galana River in Kenya.

The skeletons of individual coral animals, called polyps, fuse together and form large colonies of coral. Living polyps are found only on the surface of a reef.

Reefs are found in many different shapes and sizes. Their coral framework provides a bewildering variety of habitats for other plants and animals. Some of these habitats are exposed to surging water currents while others remain very sheltered. Among coral reef inhabitants are starfish, sea urchins, sea cucumbers and sea anemones. There are also sponges, frequently brightly coloured and sometimes massive.

Other inhabitants include sea squirts and many fish – some living there permanently while others visit occasionally to feed. But this is just the tip of the animal iceberg. Visit a reef at night or turn over a lump of dead coral and many more creatures are revealed. Molluscs – such as oysters, clams, cone shells and cowries, as

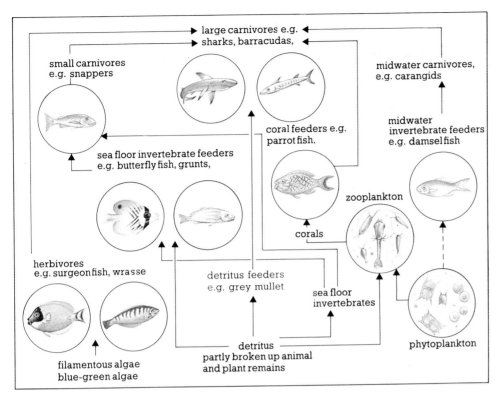

The crown-of-thorns starfish feeds on coral polyps. They have devastated living coral in many places, including much of the Great Barrier Reef in Australia.

This simplified coral reef food chain shows a variety of the colourful fishes that live in and around coral reefs, and indicates their feeding relationships.

well as crabs, shrimps and barnacles – and a host of annelid worms, live concealed in small crevices, in reef sediments or in holes in the coral rock. As many as 220 different species of animals have been found in one lump of coral alone.

What do all these animals eat? Since there is little food to 'import' from the surrounding waters, reefs are thought to be self-sufficient, producing as much food as they consume. A fundamental reason for this high productivity is the large amount of light energy available from the Sun penetrating the shallow waters. The sunlight enables a large mass of plants to grow on reefs and these plants are then available for animals to eat. Nutrients, such as nitrogen and potassium, which are not available in the waters adjacent to reefs, are effectively held and recycled on the reefs themselves.

Various types of algae are common on reefs including filamentous, or string-like, algae that often form dense mats on dead reef surfaces and also live in layers inside the coral skeleton. Many of these algae, and bacteria, are able to convert nitrogen into nitrates like the bacteria found in root nodules of legumes on land. Minute plant cells, called *zooxanthellae*, live in the tissues of corals. *Zooxanthellae* are vital to the corals because they produce energy by photosynthesis and pass this energy, in food compounds, directly to the tissues of the corals. They also remove waste products from the corals. The close relationship of corals and *zooxanthellae* may be the reason why coral reefs form only in shallow, sunlit water – the *zooxanthellae* need sunlight for photosynthesis.

Coral reef food chains are complex and there is still much to be learned about them. Phytoplankton, so crucial in many marine food chains, are generally unimportant in coral ones. Herbivores, including an astonishing number of fish, feed directly on the algae of the coral reefs. Detritus feeders also occupy an important place in reef communities. They help eliminate waste and they are themselves the main food source of zooplankton.

At the top of the coral reef food chains there are large carnivorous fish, molluscs and some kinds of crustaceans, those animals that have a hard, jointed outer skeleton. The corals themselves occupy a central position in reef food chains and are extremely versatile feeders.

Mangrove Swamps

The word mangrove has two meanings. It refers to the forest or scrub communities skirting the coastlines and estuaries of many tropical and subtropical regions. It also refers to the special group of plants that, although belonging to several different plant families, have all made remarkably similar adaptations in order to cope with the unusual conditions found in a mangrove swamp. The trees and shrubs of these areas not only have to grow and reproduce in tidal mud or fresh water and hot, drying sun but they also have to anchor themselves in soft, often waterlogged mud that is low in oxygen. Mangroves develop best in places where tides and currents allow the fine silt from rivers to settle. A mangrove forest can vary from thin, coastal strips to belts that are several kilometres wide, especially where estuaries extend far inland.

The most striking features of mangrove trees are the extraordinary root formations. These elaborate systems of buttress and prop roots, as in the *Rhizophora* species of tree, enable them to anchor themselves and feed in the shallow, oxygenated layer of surface mud.

The *Avicennia* and *Sonneratia* species of tree have conical root projections, known as pneumatophores, that grow upwards and probably help the trees to 'breathe'. All these complex root systems also speed up the accumulation of silt and debris because leaves and silt become clogged in the roots and these, in turn, accelerate the process of land building. Vivipary, the germination of the seed while still attached to the parent tree, is unique to mangroves. This type of germination probably gives the seedlings a better chance to become established in the mud.

Another feature of mangroves is the distinct zonation of the various tree species. Zones are formed parallel to the coastline. Pioneer trees, which can establish themselves and grow in mud flooded by every tide, are different from those species that are flooded less often, and different again from those reached only by exceptionally high tides.

Many different kinds of animals live in a mangrove swamp. In the tops of the trees there are birds, bats, monkeys and insects. Snails attach themselves to the leaves and barnacles to the tree trunks. Crabs, shrimps, molluscs, worms and fish live in the mud. Like the trees, these creatures are distributed in zones, according to the amount of salt concentration. Many animals, including mudskippers, build burrows in the mud while others, such as hermit and fiddler crabs, wander on the surface.

Mangrove swamps are highly productive and their importance as a source of energy for coastal food chains, which are based on detritus, is becoming increasingly apparent to ecologists. Mangroves produce large amounts of detritus – more than three tonnes dry weight per 4047 square metres each year for leaves alone has been recorded. When the leaves fall off the trees, many are trapped by the roots growing in the surface of the mud. In the mud the leaves are broken down by marine bacteria and fungi into detritus, which is eaten by worms, molluscs and crustaceans. These

The emergent roots of these tangled mangrove trees trap sediments and nutrients, and can be seen easily when the tide is low, as in this picture.

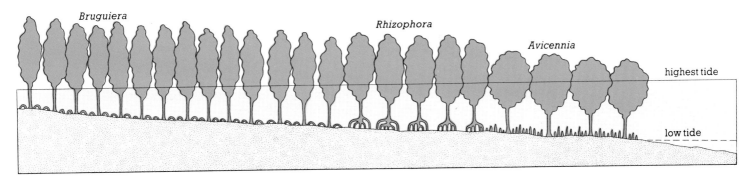

The diagram above shows a section through a mangrove swamp on the western seaboard of Malaysia, and illustrates the different zones of vegetation according to the varying tide levels.

The long root of a germinating mangrove seed (below) develops on the parent plant to prevent the seeds from being washed out to sea with the tides.

animals are eaten, in turn, by carnivores, such as fish. Fishing communities in southeast Asia catch large quantities of prawns, which feed on detritus, in the coastal waters near the mangrove swamps.

Mangroves are useful to us. They are not only an important spawning, nursery and feeding ground for many fish and shellfish, but the mangrove forests also supply wood for pulp, chipboard and firewood, as well as being a source of tannin for leather. Unfortunately, vast areas of mangroves are being dredged, filled or otherwise permanently altered to make way for roads, harbours, airports and other types of development.

Here a mudskipper fish and a mangrove snail are resting on a mangrove stilt root. Both these animals are adapted to living in and out of the water. Different species of mudskipper live in the various zones of the swamps and each species has a slightly different diet.

Tundras

Tundra, from the Lapp word for 'barren land', is the wet, arctic grassland beyond the northern coniferous forests. Tundra exists in North America, Greenland, parts of Scandinavia and the USSR. There is only a small amount of tundra in Antarctica because most of this continent is under ice and snow all the year round.

Arctic tundra regions have long, cold winters and short, cool summers. There is no light in winter near the North Pole but, during the summer, there is light night and day for up to four months depending on how near the area is to the North Pole. Because of the cold winters and short summers water in the soil is frozen the year round. This permanently frozen soil is called permafrost. A few plants are able to grow in a layer of soil above the permafrost that thaws briefly in summer. This vegetation consists mainly of small plants, including grasses, sedges, and lichens, and low shrubs, such as dwarf birch. Thawed tundra soils are badly drained and waterlogged, because of the permafrost. Decomposition of plant remains is slow in these cold, wet conditions and the vegetation often grows on mats of undecayed plants.

Although the summer season is short, plants and animals make good use of the long hours of daylight. Most tundra plants are pollinated by the wind but some, including members of the buttercup family, are

The illustration below shows a selection of low-growing tundra plants. Beneath the dwarf birches and willows there is normally a thick covering of mosses and lichens. Some plants, such as the wood cranesbill and purple saxifrage, have brightly coloured flowers that bloom in the short tundra summer.

These members of the buttercup family, called Ranunculaceae, *attract insects by focusing light and heat on to the flower centres. Surface soil and air around a growing plant are often much warmer than the air a metre or so above because soil and vegetation absorb, and disperse, some of the Sun's heat.*

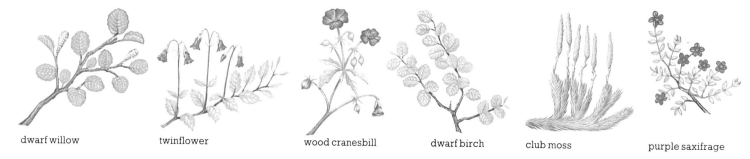

dwarf willow   twinflower   wood cranesbill   dwarf birch   club moss   purple saxifrage

Long-tailed jaegers are one of the migratory bird species that breed in the North American tundra in summer, taking advantage of the short period of abundant food supplies.

pollinated by insects and have special adaptations to attract insects. The petals of the flowers focus light and heat on to the centre of the flower where the reproductive parts are situated. These warm spots are particularly liked by flies, which move between the flowers seeking the hot places, and so act as efficient pollinators. Some tundra plants reproduce without depending on insects or wind. They send out long underground shoots that emerge as new plants.

Despite the harsh climate many animals live in the tundra. Some live there throughout the year and have adapted to combat the winter cold. Small animals, such as lemmings, survive by tunnelling under the snow. Tunnelling reduces their heat loss because snow is a good insulator. Other small animals, such as marmots, hibernate, which means that they go to sleep during cold periods. Large animals, such as caribou, reindeer and elk migrate to forests farther south during the winter. Birds, such as terns, also migrate, as do the biting flies found in large numbers in summer.

The fact that the tundra environment can only support few plants and animals means that tundra food chains are short and simple. When food supplies fluctuate, animals such as lemmings go through corresponding cycles of low and high population levels.

Temperate alpine areas such as the Alps and Rocky Mountains are often described as containing environments like tundra at high altitudes. Winter precipitation, however, is often higher there than in arctic tundra. In summer such regions are drier than the arctic because there is no permafrost. Summer temperatures are often higher and the length of the day is the same as in other temperate regions. Alpine areas are more productive than tundra. Alpine meadows, for example, are often cut for hay and grazed by domestic animals – a use that true tundra could not support.

These reindeer in the snow in Finmark, in the north of Norway on the edge of the tundra belt, eat low-growing tundra vegetation, especially lichens. Reindeer use their antlers to dig through the snow in order to find the plants on which they feed.

Deserts

We usually think of deserts as very hot sandy places with no vegetation but there are some that are very cold in winter and others, such as the Arizona desert, that have many plants. Deserts are not, in fact, characterized by their temperature but by lack of water; they are usually found in places with less than twenty five centimetres of rain a year.

There are two major types of desert. The first type includes those, such as the Sahara and Australian deserts, that lie at the edge of the subtropical regions. These deserts receive little rain because they are in an area where circulating air comes down from high in the atmosphere, absorbs any moisture in the air and carries it away to the tropics. Other deserts, including the Great Basin in the United States, are situated on the leeward side of mountain ranges and are known as rain-shadow deserts because the mountains act as a barrier to rain. Some deserts, such as the Sahara, are hot all year round but most rain-shadow deserts are hot in summer and cold in winter. The Gobi desert, for example, is often swept by freezing winds and snow.

Rainfall is rare, especially in hot deserts, but occasionally there will be a torrential downpour after a long dry period. Because the rain is so sudden and falls on dry ground it does not drain through the soil but erodes the surface to form gullies, called wadis or arroyos. These gullies funnel soil to valley floors, creating temporarily fertile areas. These fertile regions support plants that can grow rapidly.

Desert plants are adapted to living in these conditions. Many desert plants are annuals and remain as dormant seeds when it is dry. After rain, the seeds suddenly burst into life, quickly flowering and setting seed. Some plants, such as cacti, are succulents, meaning that they store water in their stems and have small, spiny leaves to reduce water loss.

Desert shrubs, such as sagebrush and saltbush, have enormous systems of roots that gather the little soil water that is available. Such shrubs also possess thick leaves that they often lose during dry periods, so preventing water loss by evaporation. Individual plants in deserts are widely spaced with bare soil between them. This spacing, which is sometimes caused by plant roots that secrete compounds toxic to other plants, means that the individual plants do not have to compete for the limited soil water and can obtain enough for their needs.

Desert animals are also well adapted to their environment. All animals, including man, produce so-called metabolic water when their bodies convert sugar into energy. Some desert animals retain this water exceptionally well. Many insects, such as tenebrionid beetles, have a thick, waxy outer skeleton that

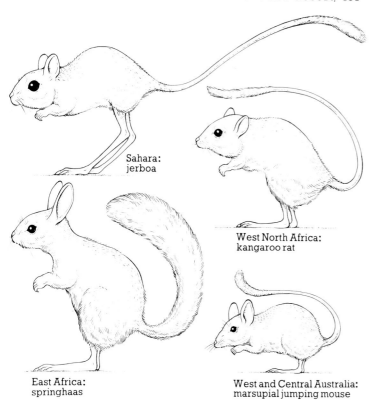

Sahara: jerboa

West North Africa: kangaroo rat

East Africa: springhaas

West and Central Australia: marsupial jumping mouse

These small rodents from different desert regions of the world are similarly adapted to their environment. They all survive partly on water produced during respiration. All have strong hind legs that make these animals speedy and agile, help them to escape predators and reduce the area of their bodies that touches the hot sand. They use their long tails like rudders to steer themselves as they leap through the air. They also have very acute hearing.

prevents water from escaping out of their bodies. Some desert mammals, such as the Saharan jerboa, live on dry seeds and do not drink. They stay in underground burrows during the day, coming out only at night, when the temperature is lower and almost the same as the temperature in their burrows. Because they are active only at night, their bodies are able to retain water that would otherwise be needed to keep them cool during the day. Such animals also excrete very concentrated urine so that little water is lost from their bodies in their waste products. Their only source of water, like that of the beetles, comes from the breakdown of sugar inside their bodies. Other animals, such as camels and road-runner birds, depend on waterholes at oases but can survive long periods without water because they can tolerate high body temperatures and the drying up of body tissues.

Cold deserts, unlike hot ones, often have an annual flush of animal life. In the Mojave desert, for example, mites and termites are often seen on the milder days of winter. They may appear for only a few days and then disappear for a whole year. For most of the time they live in deep burrows but the cool winter days allow them to surface without danger of drying up, and then they are able to mate.

Most plants cannot grow in this dry, sandy, windblown region of the Sahara desert in Algeria. Only on the most stable dunes can a few grasses take root. The Sahara is the largest desert in the world, and has an area of about 10 million square kilometres.

In the highest parts of the Atacama desert in Chile, South America, plants have to withstand cold as well as drought. At 2400 metres these Tephro cacti have a cushion-like form that minimizes frost damage.

Grasslands

Natural grasslands once covered between forty and forty-five per cent of the Earth's land surface, but many areas that were natural grasslands are now used for the cultivation of crops, such as wheat and maize. Grasslands are found in many regions of the world. In temperate areas, where rainfall is between twenty-five and seventy-five centimetres a year, grassland is the climax community. Here, it is too dry for forests and too wet for deserts to form. Temperate grasslands, such as those of the Russian steppes, the South American pampas and the North American prairies contain few trees and shrubs.

A type of grassland, known as savanna, develops in tropical areas. Although these areas have between 100 and 150 centimetres of rain a year, they also have long, dry seasons and occasional fires. Savanna contains many isolated trees and clumps of trees as well as grasses. Natural savanna is found chiefly in Africa, South America, and Australia. Grassland may also develop in places that not only have no prolonged dry season but also have high water levels or frequent, seasonal fires that prevent trees and shrubs establishing themselves. Forests once grew in parts of Illinois, Indiana and Ohio. Fires destroyed the forests, and

Although widely dispersed over the world, many of these grassland animals from Africa, Asia, North America, South America and Australia are similarly adapted to living in the same kind of environment, with its wide open spaces and dominant grass species.

	Africa	Asia	North America	South America	Australia
predators	cheetah	wolf	coyote	maned wolf	tiger cat
grazers	zebra	saiga	pronghorn	guanaco	kangaroo
burrowers	cape mole rat	bobac marmot	pocket gopher	tuco tuco	marsupial mole

North American prairies
blue grama grass

South American pampas
Dallis grass

African savanna
red oat grass

Australian savanna
spinifex

Russian steppes
crested wheat grass

Grasses are well-adapted to their environment and can tolerate a high degree of trampling, flooding, drought and, sometimes, fire. In some very dry grassland areas the leaves of grasses may roll up and form a tube to hold in moisture.

The giant ant-eater of a long snout and a very long, sticky tongue. It tears open termite mounds with the long claws on the end of its forelegs and then scoops up the termites with its tongue to eat them. The ant-eater's furry coat helps to protect it from the bites of insects.

prairies took their place.

Different types of grasses grow in different regions of grassland. Where rainfall is high, large tall grasses such as pampas grass (which grows to about 2.4 metres) can flourish. In drier areas shorter grasses (between 0.6 and 1.2 metres high) are dominant. Low-growing bunch grasses, such as the grama grass of the United States, grow in very dry areas, near deserts.

Some grasses grow in different seasons, according to the climate of the region that they inhabit. Some, especially those of temperate regions, such as cocksfoot, grow during the spring and autumn, remaining green throughout the winter. In the summer they are semi-dormant, and grow little. Other grasses, particularly those of tropical grasslands, grow in late spring and summer, but not in the autumn and winter. Many grassland animals have adapted to these patterns of growth that vary with the seasons. Some antelopes, for example, migrate, while ground squirrels hibernate during the periods of little grass growth.

Many grassland animals are easily seen because there is little vegetation cover. In order to avoid predators some animals, including antelopes, run very fast and herd together. Others, such as rodents, burrow among grass plants to avoid detection. Predators, such as coyotes and hawks, are important in grassland communities because they keep the number of herbivores, particularly rodents, at a level that can be supported by the plants. If the predators are removed the numbers of their prey increase dramatically.

Grazing animals are responsible for the number and types of grassland plants in a particular area. Grasses often make up as little as twenty per cent of the number of plant species, but are dominant because of the way they grow. Too much grazing reduces the number of plant species because some will be completely eaten away. Too little grazing allows some plants to grow at the expense of others. The stronger plants cause the death of the weaker ones by overshadowing them and competing with them for water and food.

Forests

Forests are complex ecosystems dominated by trees, which provide protection for other forms of life against the full impact of sun, wind, rain and snow. Forests can develop wherever the average temperature exceeds 10°C in the warmest months and where the annual rainfall is greater than seventy centimetres. There are three basic kinds of forests, determined by the type of soil and by the prevailing climate of the region: coniferous, deciduous and rain forests.

Coniferous forests grow in acidic soils, and most conifers are evergreen, which means that they have a covering of leaves all the year round. Conifers are able to withstand drought (which can be produced by either dry or frozen soil) and can take up water as soon as it is available – either from rainfall or from thawed soil. Although coniferous forests are extensive in the northern hemisphere (the so-called 'boreal' forests), where the winters are long and cold, they also grow in Mediterranean climates in both Europe and North America, in mountainous regions, and in warm, temperate regions such as the subtropical areas of the south-eastern United States. Coniferous trees are found in South America, Australia and New Zealand, but there are no forests dominated by conifers in the southern hemisphere.

Most deciduous forests grow in brown earths and the trees have leaves that they lose in autumn. They grow in temperate regions with marked seasons, such as the middle latitudes of Europe, the USSR, eastern North America and eastern Asia.

Rain forests, in which tall trees dominate, grow in tropical, warm temperate and cool temperate climates. Although these regions vary in temperature, they all have a high rainfall that produces prolific vegetation. Tropical rain forests are evergreen or semi-evergreen. Warm, temperate forests contain both evergreen and deciduous trees, and sometimes conifers. They are found in both the northern and southern hemispheres whereas cool, temperate rain forests grow in the southern hemisphere. Here the trees are broad-leaved evergreens, sometimes with conifers or deciduous trees.

This diagram illustrates the layers of vegetation found in a coniferous, deciduous and rain forest, and shows how the complexity of the layers varies from one type of forest to another, especially in the abundance of undergrowth and climbing plants.

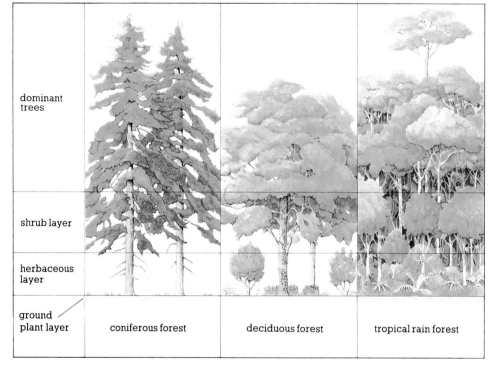

The trees in this coniferous forest in Bavaria are very tall and there is little undergrowth. Conifers such as these have foliage all the year round.

Ground-layer plants in deciduous forests, such as this beech forest in Bavaria, are often low-growing and flower in early spring before the leaf canopy develops.

Along their margins, forests merge into other zones. Boreal forests, for example, gradually change to tundra or deciduous forests. Deciduous forests graduate into prairies and tropical rain forests into savanna.

The vegetation of forest communities grows in several different layers, which are known as ground, herbaceous, shrub and tree layers. The pattern varies according to the types of tree. Coniferous forests have the simplest structure, with a continuous tree layer and a ground layer of mosses, liverworts and lichens. Deciduous forests contain more shrubs and flowering plants as well as climbers, such as ivy. These various layers provide very different habitats for animal communities.

The most complex structure is found in tropical rain forests. These forests have three tree layers, as well as a sparse shrub layer, and a herbaceous layer with tall ferns and many woody climbing plants and epiphytes, those plants that grow on trees rather than in the ground. Such forests support rich and varied animal communities.

Forest animals are well adapted to their habitat. Sharp claws and long tails enable tree-living animals to grip and cling on to the branches. Many animals, including squirrels, make their homes in the trees.

Although forests are the most efficient communities on the land for the production and storage of energy, much of it is 'locked up' in the trees themselves and is not available to the rest of the community except when released as fruit or seeds. Food in the forest is therefore scarce and, except in the tropics, animal populations are relatively low.

Coniferous and Deciduous Forests

Travelling south from the frozen expanse of the arctic tundra, the first true forests appear in the region known as the taiga. These are the coniferous, mostly evergreen forests that are dominated by pines, spruces and larches with a sprinkling of birch trees. Coniferous forests stretch in a band across northern Europe, into Siberia, and occupy large areas of the North American continent. Conifers also grow high up on mountains in the Asian and American tropics.

Conifers are so-named because their seeds are protected by woody structures called cones. They are older in evolutionary time than other trees, and depend on wind and air currents rather than insects for pollination. Animals and birds that bury seeds for food stores also play an important part in seed dispersal.

Conifers are among the tallest trees that we know, their straight trunks soaring up to a dense canopy of dark green, needle-like leaves shading a gloomy forest floor. Little else grows here, except for patches of herbaceous plants and a ground layer of mosses, liverworts and lichens, inhabited by ants and slugs.

Rain and snowfall are relatively low – perhaps twenty-five to fifty centimetres a year. The ground is often frozen for half the year, making water unavailable to the tree roots. The trees can endure these drought-like conditions because their leaves have a small surface and thick skin that reduce the loss of water.

Evergreen leaves do fall from time to time, creating a litter layer on the forest floor. The fallen leaves are replaced by new leaves, so the trees never become bare. Conifers can thus manufacture and store food all the year round, making the best use of the available

In Oaxaca state in Mexico there are transitional areas of semi-desert where deciduous trees grow alongside cacti. In this photograph the leaves on the trees are changing colour because of the approaching dry season. The conditions here are dry enough for cacti but also wet enough for deciduous forest.

light from the low-angled northern Sun.

Only about ten per cent of sunlight reaches the forest floor. Beneath the litter layer of dead leaves, fallen branches and other dead plant material, the soil is sandy, quick-draining, and acidic. The top layer is relatively poor in minerals as these are dissolved away by drainage, and settle above the bedrock layer. Bacteria cannot flourish under these conditions and therefore the production of humus is slow, helped mainly by fungi. Earthworms are scarce or absent, so the soil layers remain undisturbed and poorly aerated.

Deciduous trees, which lose their leaves in the autumn, become mixed with conifers in more temperate climates. Most of northern Europe is in the coniferous-deciduous forest belt, and there are large areas of mixed forest in eastern North America. Such forests, characterized by oak, beech and pines, are not necessarily transitional between the forest types but may be climax communities in their own right.

This diagram shows a European coniferous forest in all seasons, and a deciduous forest in summer and winter. The inset drawings illustrate the plants and shrubs of the forest floor and a selection of animals and insects that might be found on, or just beneath, the ground.

Deciduous forests of broad-leaved, flowering trees such as oaks, beeches, birches, aspens, elms, sycamores and maples are found in North America, Europe and China, where the winters are cold and the summers are warm and wet. The large, delicate leaves of these trees make the maximum use of the spring and summer Sun for photosynthesis. Water losses are minimal in winter, when the trees have lost their leaves. Such trees depend on wind and insects for pollination, and on wind, birds and animals for seed dispersal.

In full leaf, the tree canopy allows about five per cent of the sunlight to reach the forest floor and, when leafless, fifty to ninety per cent. Shrubs, grasses, ferns and flowering plants grow here, and woodlice, snails, slugs, centipedes, spiders and beetles live among them. The typical brown earth soil is fertile, rich in the bacteria that break down the litter layer into humus, and is well mixed up and oxygenated by earthworms.

Forests are the most efficient and productive of all ecosystems, yet only about five per cent of net primary production in a temperate, deciduous forest goes to animals. Most nutrient recycling takes place through decomposition. Nevertheless, these forests support a diversity of wildlife including birds, rodents and top carnivores such as the fox, owl, lynx and wild cat.

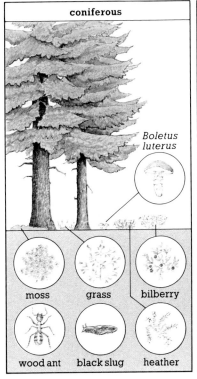

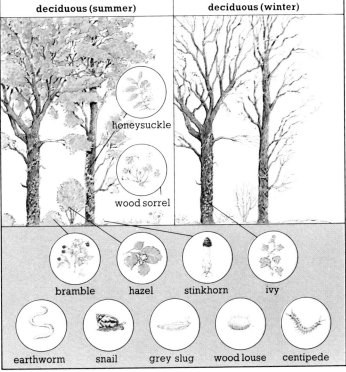

The photograph above shows the seeds inside a pine cone protected by the outer scales. Male and female cones are often found on separate trees.

Tropical Rain Forests

Tropical rain forests are the most complex ecosystems in the world. They are found at low altitudes near the Equator where the rainfall is between 200 and 225 centimetres or more a year. Major rain forests are situated in central South America, the Panamanian isthmus, central and West Africa, Madagascar, and Southeast Asia. Although rain forests may have seasonal patterns of rainfall, the temperature varies little from the warmest to the coldest months. The Panamanian rain forest, for example, has most rain during its warmest period (April to July) though rain falls during the cool months (November to January) as well.

All rain forests have the same structure even though they may contain different types of plants and animals. Rain forests can be divided into several layers, each of which is almost an independent area. The top layer consists of the tops of the highest trees, known as emergents, because they stick up above the layer of trees beneath them. The next layer is about twenty-four to thirty metres from the ground and consists of an almost continuous canopy of evergreen trees with broad leaves. Where plenty of light can penetrate the canopy, the bottom layers of the forest consist of shrubs and an undergrowth of ferns and herbs. Where little light penetrates there is often no shrub layer and the undergrowth is sparse. Within the forest there are many creeping plants, such as vines, that climb up tree trunks and many different kinds of lichens and mosses that coat the branches and trunks of the trees.

Despite the almost continual heat and heavy rain, many plants and animals behave as if there were distinct seasons. Many birds, for example, breed according to the patterns of rainfall and some emergent trees lose their leaves during drier periods, particularly when there is less than five centimetres of rain a month.

Tropical rain forests are very productive. This productivity is due to rapid nutrient cycling. Dead plants and animals are rapidly decomposed and the nutrients released by this action are quickly absorbed by plants.

This cloud forest in the Rancho Grande, Venezuela, has a broken canopy that allows sunlight to filter through, enabling smaller plants to grow.

The surface of the soil contains many tree roots that absorb nutrients before the nutrients reach the soil lower down. Consequently, this lower soil is often infertile. Some root systems are closely linked with fungi, forming complex root webs in the leaf litter.

Animals in rain forests often live in a particular forest layer, mostly in the canopy. Many of the brightly coloured birds, such as the toucans and birds of paradise, fly around in the canopy, eating the fruits on the trees. Others, such as hummingbirds and sunbirds, feed from nectar in the flowers, often from the epiphytes that grow on the trees. Many mammals also live in trees, such as flying squirrels and flying foxes. Such animals are adapted for life in the trees – flying foxes can glide and tree sloths have long, gripping claws to

hold on to the branches. Other canopy animals include butterflies, flying chameleons, and flying frogs. Many animals also live on the ground, such as snakes and leaf-cutter ants. Most of the activity in a rain forest takes place at dawn, dusk, and during the night when animals such as bats, tree frogs and slow lemurs emerge, and make the forest noisy.

Different types of rain forest grow in high places in the tropics. With increasing height there are fewer emergent trees and a more broken canopy. At about 1520 metres small trees are found growing up into the clouds. Above 1820 metres the forest vegetation looks like that of temperate coniferous forests, except in Africa, where there are no conifers.

Many roots of tropical trees are close to, or lie on, the surface of the soil where they can obtain nutrients from decaying plant material. These Rajah Brooke's birdwing butterflies have flown down from the canopy to feed from the plant litter.

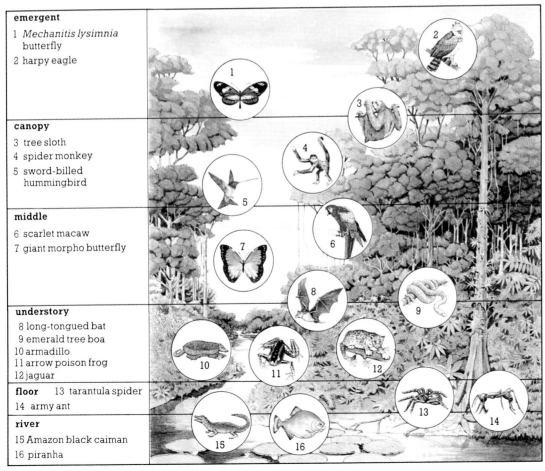

emergent
1 *Mechanitis lysimnia* butterfly
2 harpy eagle

canopy
3 tree sloth
4 spider monkey
5 sword-billed hummingbird

middle
6 scarlet macaw
7 giant morpho butterfly

understory
8 long-tongued bat
9 emerald tree boa
10 armadillo
11 arrow poison frog
12 jaguar

floor   13 tarantula spider
14 army ant

river
15 Amazon black caiman
16 piranha

This diagram of tropical rain forest in South America shows a selection of the kinds of animals that live in the various forest layers.

Ecology and People

Human beings have tried to understand the ways of the living world since the dawn of civilization. For early people, much of this interest was based on having to hunt for food and clothing, while the first cultivators needed to know how to grow as much food as possible. The knowledge that such people employed was ecological. It was essential for hunters to know the habitats and migration routes of their prey. Early cultivators discovered, by trial and error, the best crops for their soils and the best way of growing their crops in order to limit damage by pests.

Many of the traditions of these early hunters and cultivators persist. Lapps, for example, follow migratory reindeer, and many peasant farmers grow their crops in small, mixed blocks to reduce pest harm. Until recently the nomadic herders of Africa followed seasonal routes with their herds, taking care not to use up too much of the vegetation. Similarly, the Tibbu people of Libya wander from pasture to pasture with their herds and flocks, and may stay in one place for as long as two or three months before moving on. Many of the traditional routes and pastures of wandering nomadic tribes are now impassable due to wars or settlement policies. The restrictions imposed on these peoples is one of the causes of the spread of deserts in Africa, because such tribes can only graze their animals in one area and so use up all the vegetation growing there.

As the human population and standards of living for richer nations have increased, so the quantities of all materials taken from nature have risen. Not all countries have benefited equally from this harvesting of the Earth's resources. Apart from the problems of starvation or malnutrition in poor countries, over-consumption of resources also creates pollution in both rich and poor areas, and can destroy whole ecosystems.

Many ecologists and economists now realize that mankind is not divorced from the rest of the natural world but is part of it. We now understand the effects that we are having on ourselves and other plants and animals by our damage of nutrient and mineral cycles and our over-using of resources such as the soil and forests.

Thomas Malthus, an eighteenth-century British economist, was one of the first people to realize that continuous population growth can create havoc with the environment and impoverish natural resources by using them up too quickly. Malthus published his famous treatise on this theme, *Essay on Population,* in 1798. Charles Darwin, the famous nineteenth-century British naturalist, was one of the first modern ecological

This portrait of Charles Darwin (1809-1882) shows the British naturalist in his old age. Darwin formulated the theory of evolution by natural selection. As a young man, he spent five years working on board HMS Beagle *and visited, among other places, the Galápagos Islands in the Pacific. This five-year cruise was crucial for the development of his concepts about the origins of species.*

A Dinka tribe in the Sudan tend a herd of long-horned cattle. The Dinka live in harmony with their environment. In the dry season they set up camp with their cattle near the Nile River and, in the rainy season, they move to higher wooded ground, where they cultivate crops.

thinkers. In Darwin's book, commonly known as *The Origin of Species,* published in 1859, he proposed that different species had evolved by the process of natural selection. Individual species, he wrote, are adapted to their environment and evolve in response to changes in that environment. If, however, the changes are very rapid, species may become extinct if they cannot adapt at the same rate.

This process of selection and adaptation helps us to understand why sudden environmental changes caused by human beings often result in the destruction of plants and animals and, sometimes, whole ecosystems. Natural changes in environments take place over thousands of years and most species can adapt gradually, generation by generation, to the slowly changing conditions. People, however, can completely change an environment within only a few years.

The traditional form of rain forest cultivation, still practised in a few places, is to clear a small area and cultivate it until yields decline. A new area is then cleared leaving the old one to return to forest vegetation.

Soil Erosion and Conservation

The soil is one of the Earth's most vital assets; without it there would be no plants on land and we would not be able to grow crops for food. The forming of soil is a slow process – it takes thousands of years for nature to build layers of soil rich enough for farming.

Wind and water both help to create soil, but they can also destroy it. Growing plants normally bind the soil together with their roots so that any water trickling through does not wash away the soil particles. The leaves and stems of plants also prevent damage to the soil by wind, and protect it from heavy rain. Soil that has no vegetation may be in danger. It may be washed away and end up as silt in flood plains, deltas or even in the sea, or it may be blown away by the wind.

People are also responsible for eroding the soil. This is not a new problem. The Sumerian civilization that flourished from roughly 3100 to 2000 BC on the flood

Severe soil erosion around Deer River in Colorado, USA, was caused by water running down the surface of hillsides, forming gullies, and removing all the topsoil, as shown in the photograph on the right.

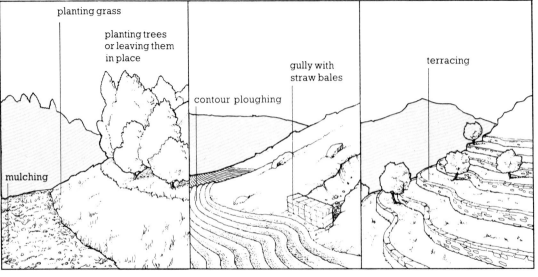

This three-part diagram shows various methods of soil conservation on both flat land and on hillsides where soil can be easily blown or washed away.

plains between the Tigris and Euphrates Rivers (now part of Iraq) caused a great deal of erosion. Silt from their farms washed into The Gulf, adding 290 kilometres to the coastline.

Farming always changes the soil to some extent because, when the soil is left without vegetation between crops, wind and water may erode it. Allowing animals to graze too much causes plants to be eaten away, also leading to bare soils and so to the likelihood of erosion.

Today, the most severe areas of soil erosion are in the tropics and subtropics where there is heavy rainfall. Cutting down forests in the Himalayan foothills is removing much of the vegetation from the watershed (the area of drainage) of the Ganges River. Eroded soil is being carried down to the mouth of the Ganges, forming an island in the Bay of Bengal. In the 1930s large areas of the plains in the United States were eroded by the wind after they had been ploughed up and used to grow wheat. This erosion raised clouds of swirling soil particles into the air, creating huge dust bowls. In some places as much as twenty-five centimetres of soil were blown away. American farmlands are still losing topsoil at an alarming rate.

Methods of conserving soil are simple to apply but are sometimes unpopular because they result in some loss of land that could have been used for cultivation. Erosion by water, for example, is usually worst on steep hillsides. This type of erosion can be prevented by terracing (building low stone walls step by step up the hillside) and contour ploughing (ploughing around in line with the shape of the hill). These practices reduce the speed of water flowing downwards through the soil. Existing eroded gullies can be dammed and permanent vegetation, such as bushes and small trees, can be planted on badly affected areas of soil. Mulching (spreading a covering on the ground, such as straw) also restricts erosion by water where there are no plants or trees.

Erosion by wind can be prevented by planting belts of vegetation to give shelter. Strip farming (growing crops in narrow strips) helps to prevent erosion by both wind and water because, by cultivating parts of the land at different times, only a small proportion of the soil is left bare at any one time.

Sometimes the actual structure of the soil is damaged. Trampling by grazing animals and the use of very heavy machinery can press down the soil so hard that the different layers are squeezed together. This forms hardpans, which prevent plants from growing because they cannot obtain water and nutrients.

It is possible to improve areas of soil that are compacted or where humus is thin. Ploughing the subsoil may help pressed down areas to recover. Similarly, plant and animal waste products can be added to the soil to make humus. Unfortunately, both methods are expensive and therefore only used when essential.

This dust storm in Texas, USA, was caused by the wind blowing dry soil from areas with no vegetation cover. Texas was affected by a severe drought in 1980 and much ground baked hard and then cracked.

Mankind and the Forests

Virgin forests are rare. Probably the only untouched forests that remain are parts of tropical rain forests, some northern coniferous forests, areas of the redwood forests of North America and the southern beech forests of South America.

People have been clearing forests for their crops since farming first began. In central and western Europe, for example, forests were being cleared in Neolithic times (about 4000 to 2500 BC) and the present woodland pattern was probably already established by the fifteenth century. In North America the rate of woodland removal has been more rapid. Since colonization by Europeans, forests have declined from around 295 million hectares to about 400 000 hectares. The tropical rain forests are now being cleared at an alarming rate. At least forty hectares are being felled every minute. Not only have the forests been removed to make way for crops but also the products of the forest (in the past, timber for building and firewood and, more recently, pulp for papermaking) have always been in demand.

In Europe and some parts of Asia there is a long tradition of forest management. Managing forests means, traditionally, encouraging particular types of trees for good quality timber used for building and for wood to make tools and fencing. This method relied on the capacity of the forest to grow again. There was little planting and the system worked only if a small proportion of the total forest was cut down. The combination of clearing, forest management and hunting caused many species of animals to become extinct. In Britain the brown bear disappeared in the tenth century and the wolf became extinct in the eighteenth century.

Until recently the main technique of forest management was to plant large areas with conifers, often on the sites of native woodland. These planted forests suffer from major outbreaks of pests, which are the result of growing monocultures (cultivating one species only). Monocultures often have little resistance to certain pests. The pine looper moth, for example, can destroy large areas of trees if they are all of the species that it prefers. Another problem of monocultures – and of some of the native coniferous forests in North America – is that clearing large areas at one time may result in soil erosion.

Rain forests, now in danger of destruction, contain almost half the plant species of the whole world and about eighty-five per cent of the insect species of the world. Several rain forest species are commercially successful, such as the rubber tree. New uses are constantly being found for other products of rain forest plants and animals, particularly for medicine, new crops, and for pest control. Unfortunately, many rain forest species are in danger of extinction because of extensive clearance, and before we have discovered whether they may be useful for us. Many ecologists also fear that the removal of rain forests may affect the climate of the earth by altering the levels of oxygen and carbon dioxide in the atmosphere.

Felshamhall Wood in Suffolk has been managed since the thirteenth century. The trees shown here are oak and birch, and the stacks of chopped wood are alder and ash.

Rain forests are often felled for short-term financial gain. Areas of the Amazonian rain forests are now being cleared for growing grass to feed cattle, supplying the United States, and other western nations, with beef. Because the soils of rain forests are infertile, grass production drops after a few years, leading to erosion and hardened soils. Finally, the area may be abandoned. So much forest has been removed in some places that it is almost impossible for it to grow again by itself because there are no colonizing plants left. Rain forests can be saved by proper ecological management but it is expensive to apply this knowledge. Nevertheless, the long-term benefits of looking after the forests correctly would undoubtedly be worthwhile.

This rubber plantation in Malaysia is an example of an export crop that is beneficial to the natural environment. The rubber trees provide foliage cover over the soil so that the rain does not wash away the soil, and the root systems and fungi help to bind the soil together, thus preventing laterizatin. The rubber plantation in fact mimics the natural rain forest that it has replaced.

This bare hillside in northern Thailand has been caused by extensive deforestation. A small remnant of the original forest remains on the top of the hill in the background.

Mankind and the Seas

For thousands of years people have been searching the seas for food. Before reliable methods of navigation were invented the only seafoods that could be safely harvested were those on the beaches and rocky shores, such as shellfish and seaweeds, and fish that could be caught from small boats near land. With the development of fishing techniques and methods of preserving fresh fish, it was possible to fish farther from the shores, catching herring and pilchards, for example, on the continental shelves, the shallow areas bordering the coasts. Fishing in deep waters for fish such as the Atlantic cod has only developed on a large scale within the last hundred years.

Today, people fish all over the oceans but the best fisheries in the world are those of continental shelves and in the regions of cold oceanic currents. These areas contain plenty of nutrients for the plant plankton that attracts fish.

Few fish are, in fact, of commercial interest. Almost half the world's fish harvest consists of only six species: Peruvian anchovies, herring, cod and mackerel from the Atlantic, Alaskan pollack, and South African pilchards. Most of the commercial fish feed on plankton though some, like tuna, are carnivorous. Surprisingly, almost half of the fish caught are not eaten by humans but are turned into fertilizers and food for poultry.

Recently many of the world's fisheries have been yielding fewer and fewer fish. This is because, in the past, too many fish have been caught and the ecology of the seas has not been properly understood. Not only has previous overfishing reduced the amount of fish being caught today, but it has also damaged or altered marine food chains. The sardine and anchovy, for example, are both plankton feeders that compete with each other for food. Overfishing of the sardine in the

This puffin is eating sand eels – fish that are lower in nutrients than the puffin's usual food, herrings, which are in short supply because of overfishing. Puffin colonies are found on the Arctic coasts, and as far south as Brittany and northern California.

Oyster cultivation is popular in France. This oyster farm contains hundreds of oysters that grow rapidly in the waters of estuaries, which are rich in plankton.

Pacific in the early 1940s caused their numbers to fall, and the sardine has now been replaced by the anchovy as the main plankton feeder along the Californian coast.

Attempts are now being made to manage fish stocks by applying ecological knowledge. Unfortunately, these attempts are often hampered by international wrangles, such as the arguments among the European nations about fishing boundaries and the quantities of fish that can be taken from the North Sea, the English Channel, and the North Atlantic.

Too much fishing has also affected the populations of secondary and higher-level consumers, such as seals and seabirds. One of the largest puffin colonies in the world, off the coast of Norway, produced no young in 1980. Their main sources of food – small herring – have been so exploited that they had to switch to fish with a lower food value, such as sand-eels, in order to feed their chicks. Sadly, the chicks died from starvation.

Marine mammals have also been exploited by human beings and often hunted to extinction. Steller's sea cow, for example, became extinct in the 1700s because of the activities of Russian fur trappers and sealers. The sea cow was originally found in large numbers in the North Pacific, with populations of over two million. Several species of whale, such as the giant blue whale, are also near extinction. There is hope, however, that the whales may increase, partly due to international bans on hunting blue whales through the intervention of the International Whaling Commission.

Fish farming is now gaining attention. For example, mussels and oysters that live in estuaries and on rocky shores are a great delicacy and very good nutritionally, but their numbers are often limited by a lack of places for them to settle on. Their numbers can be increased simply by setting out rafts and posts, on which their larvae can settle and grow to maturity.

The food chain of the young and adult herring, shows its feeding relationships with other sea organisms. Overfishing of the herring causes alterations in feeding habits and can profoundly affect food chains, upsetting their balance, and causing other organisms to decrease in number. The common herring is found in the temperate and cold waters of the North Atlantic, where it grows to about thirty centimetres in length. The herring family also includes other species, such as the anchovy and sardine.

Deckhands on a whaling ship, above, are stripping the blubber, or fat, from a whale. They will boil down the fat to make oil. The meat may be canned for human or animal consumption.

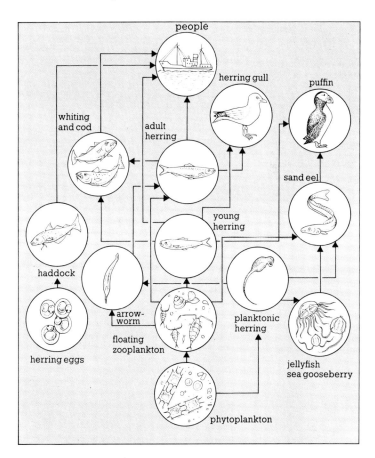

Conserving Wildlife

Throughout the history of life on Earth some plant and animal species have thrived while others have become extinct. Some plants and animals died out because their habitats altered, but most have disappeared because of human activity. Archaeological evidence suggests that during the Pleistocene period hunters were responsible for the extermination of at least thirty per cent of Africa's large mammals and seventy per cent of the mammals in North America. Until the recent past extinctions have not been well documented but, within the last two hundred years, at least fifty-three species of birds, including the dodo, and seventy-seven species of mammals have vanished in the world. Hunting animals for their fur and feathers has caused some of these extinctions. Occasionally, animals are hunted purely for amusement or out of boredom; the iguauas of Baltra, one of the Galápagos Islands, were wiped out in this way during World War II.

Changing habitats as a result of farming, however, is the greatest cause of extinctions. The widespread draining of the East Anglian fens during the 1850s, for example, changed this habitat from wet marshland to dry land, so that it was no longer suitable to support the Great Water Dock plant. As this plant became scarce the numbers of Large Copper butterflies, which fed on Great Water Docks, declined and finally became extinct because of the unchecked activities of butterfly collectors.

As the human population increases more land becomes intensively cultivated for food. More and more plants and animals are threatened by reduction in numbers or possible extinction because their natural habitats vanish. It is now estimated that about 25 000 different plant species are threatened with extinction. Many ecologists now realize that conservation is essential for the continued survival of the human race. Many of our crops, for example, are unstable and liable to sudden outbreaks of pests and disease. Such events are extremely rare in natural communities, those areas undisturbed by people. Usually pests and diseases of crops are treated with insecticide and fungicides. We now know that there are predators that can prevent such outbreaks in the natural communities from which

The dodo was a flightless bird that lived on the island of Mauritius in the Indian Ocean. It became extinct about 1680 due to hunting and the introduction of pigs on to the island, which disturbed the birds' nests.

the pests originate. Such communities are therefore capable of providing natural agents for pest control. Natural communities also provide us with the opportunity of studying ecosystems, and the knowledge gained from this can be used to develop better methods of food production, using nature's own organization as a model.

All cultivated crops and domesticated animals had wild ancestors, many of which still exist. Others, such as the aurochs, which were the ancestors of European cattle, are extinct. Cultivated crops are now so inbred that there is little material for plant breeders to select desirable qualities for new breeds. Wild varieties are not so inbred and can provide resistance to insects and viruses in new crops that are bred from them.

Wild animals are also often able to resist diseases that affect domesticated animals. On the grasslands of Africa, for example, European cattle often succumb to local diseases against which they have little or no immunity. Some farmers are now planning to raise zebra and antelope instead, whose meat is perfectly edible.

These animals, native to the grasslands, are naturally immune to grassland diseases.

In the past, conservationists have often concentrated on saving a single species. It is now realized that wildlife conservation should be based on preserving entire habitats and ecosystems. This kind of conservation not only protects individual species but it also preserves new material for cultivation and ensures the existence of species that can control pests. One example of the ecosystem approach to conservation is Operation Tiger, launched in 1973, which aims at preserving not only the rare Bengal tiger but also all its habitats.

The Californian condor, once common in the mountains of North America, is now threatened with extinction. There are only about forty birds left in southern California, but protection and monitoring of its breeding habits may ensure its survival. Condors are the largest birds of prey, with a wingspan of up to three metres. They normally eat carrion but will attack living animals as large as a fawn.

Tapirs are found only in South America and Southeast Asia. This rare Malaysian tapir is distinguished from its American relative by its black and white markings. Tapirs are timid, water-loving, jungle herbivores that emerge at night. They have long, flexible snouts and stand about one metre high when adult.

Desert Increase

More than one-third of the world's land is desert or near-desert, and this area is steadily increasing. The problem is ancient but was drawn to the world's attention in the mid-1970s when the herders of the Sahel and Sudan savanna regions of Africa faced starvation after six years of drought. During this time the rains had failed and, as a result, the grasslands could not support livestock. About 100 000 people died of starvation. Much of the Sahel savanna is now desert – an extension of the Sahara – created by the combination of drought and overgrazing.

Most of the world's deserts are of relatively recent origin, and have arisen because of changing climate and human activities. The Thar desert in India, for example, is estimated to be only one thousand years old. About two thousand years ago Alexander the Great marched his army through this region, which was then virgin forest. Similarly, much of northern Africa used to be fertile and supplied the Roman Empire with grain, olives, dates and grapes. Today it is desert and the ancient city of Carthage, once a thriving agricultural centre, is surrounded by unproductive land.

The climatic changes that have caused deserts to form are well documented, particularly in Africa. From the last glaciation (about 100 000 to 10 000 years ago) to about two thousand years ago, much of the northern Sahara was cooler and wetter than it is now. Only seven thousand years ago hippopotamus and other swamp animals were living 640 kilometres north of Timbuktu. Elephants could be found in the Atlas mountains among forest vegetation along the shores of Morocco only two thousand years ago. Since then the Sahara has become steadily drier and hotter and the plants and animals fewer in number.

Some ancient cultures were responsible for creating deserts. The agricultural practices in some parts of North Africa, for example, during the Roman era (about 500 BC to about AD 410) were largely based on complex systems of irrigation. Most of the food that was produced came from harvested plants. The land was not fertilized adequately and so crop yields went down, particularly when the Roman Empire began to decline. Eventually the irrigation systems were neglected and the Roman farmers were replaced by nomads, who overgrazed the land, leading to the formation of desert on what was once fertile ground.

Much of the recent southward extension of the Sahara is entirely due to large-scale changes in vegetation. In Somalia, for example, savanna was created from forest only two thousand years ago. The original clearance was made by farmers who were replaced by wandering herdsmen when the soils became infertile after continuous harvesting. It is, however, only recently that the savanna has been replaced by desert and semi-desert in these regions. Until the 1930s oryx, gazelles and lions were common on the plains. Human populations increased, leading to an increase in the number of their grazing animals. The overgrazing inevitably led to the formation of desert.

In overgrazed areas the animals remove perennial grasses and eat young seedlings, preventing new vegetation from taking root and sending out new shoots. In the dry season shrubs are often cut to provide fodder for animals, and trees are often removed to provide firewood. These processes lead to bare soil that is easily eroded by wind and water, further

These skeletons of cattle were found around a dried-up water hole in Niger during the Sahel drought.

Sand drifts over an old telegraph station at Eucla on the Nullarbor Plain in Western Australia.

This cave painting is in the southeastern mountains of the Algerian desert. The painting dates from about the 4th to the 2nd millenium BC, and depicts cattle herders with their animals. Such paintings are evidence that this area was once fertile.

reducing the capacity of the area to support plants and animals. Bare soil also reflects the sunlight, warming the atmosphere and reducing rainfall, while dust raised by the animals often rises into the atmosphere, preventing cloud formation. All these effects can be reversed by ecological knowledge: reduction in the numbers of grazing animals and in the amount of trees cut down, and the use of soil conservation techniques on areas of bare soil.

The Blooming of the Deserts

Although people can cause deserts to increase, much effort is also being made to reclaim some desert regions. Areas that were once barren are now producing crops, contributing to the welfare of desert peoples.

The traditionally fertile places in the desert are oases. Nearly all of these depend on water from springs and wells although some, like those of the Hoggar and Tibesti mountains in the Sahara, receive rain, which falls on high desert regions. The major crops of oases, particularly the Sahara, are date palms. Citrus fruits, such as lemons, as well as grapevines, apricots, vegetables and fodder for animals, also grow in oases. Farmers have recently expanded some oases by drawing up water from underground. The area that is cultivated around Tunisian oases, for example, has doubled within the last fifty years.

Unfortunately the water that supplies oases is limited. In some cases the water levels are dropping and some schemes to supply oases with water rely on a source that is not being replaced. In Baja California, farming projects started in the 1960s are being based on wells that may last only fifteen years.

Irrigation is the best system for reclaiming the driest parts of deserts, especially if the water supply comes from outside the desert. Such schemes can, however, create problems – both in the irrigated zone and in the areas from which the water is taken. Water evaporates quickly from the soil in desert areas and mineral salts, such as chlorides, are carried up into the top layers of the soil by the evaporating water. Although some crops, such as barley and alfalfa, can grow in very salty soil, most cannot. In order to prevent salts reaching these high levels in the topsoil more water has to be added to trickle downwards.

This photograph shows how sprays are used to irrigate the desert in Colorado, USA.

A large area of the Colorado desert is now being irrigated by sprinklers rather than by the more traditional method of using ditches filled with water. Not only does sprinkler irrigation reduce the amount of salt in the soil but it also helps to stop evaporation from the soil because the water falls with the force of rain, counteracting the upward movement of water through the soil.

In Egypt the Aswan High Dam on the Nile has created new farm land. Unfortunately, this dam is blocking sediments that normally flow to the delta at the river's mouth, causing the Nile's delta to shrink and become full of salt. The sardine fishery in the delta has been virtually destroyed because the flow of nutrients from the river is no longer enough to support sardine life, and the cultivated land has lost much of its farming potential because of erosion and salt deposits.

Some deserts receive enough rainfall to support vegetation, but the soil is too unstable for plants to take root. Techniques for stabilizing sand dunes can convert these deserts to farm land. These techniques have been used in the semi-arid pampas region of Argentina. There, dunes were first smoothed down with tractors and then planted with grasses. To prevent erosion by the wind, a thin layer of straw was ploughed into the dune. Within three months, vegetation was growing on the dune. In parts of Africa, dunes have been planted with poplars, acacias and eucalyptus – trees that take root in the dunes and stabilize them, allowing other plants to gain a foothold. These stabilizing techniques have brought life to the desert at only a small cost.

The Nile valley in Egypt is naturally productive because of the sediment deposited by the Nile River. Water from the Nile is also used for irrigation. In the photograph at the top, a farmer spreads fertilizers before planting a potato crop.

The lush palms around the standing water of an oasis in Niger, in the south Sahara, contrast sharply with the dry, plantless desert beyond. Some oases have recently been expanded by drawing up the underground water to irrigate the surrounding land.

The Growth of Human Populations

The human race is now experiencing its third population explosion. Each rise in population has followed a cultural advance that has allowed people to expand their ecological niche.

The first rise in population was the result of two developments: the ability to control fire and the ability to make weapons and tools. These skills enabled humans to keep themselves warm, to hunt a wider range of animals for food, to build shelters and to make clothing. They could then exploit new environments that were previously unpopulated.

The second rise in population was due to the development of stable agriculture on uplands and fertile river plains about 10 000 to 12 000 years ago, and to the domestication of wild animals. It was then possible to maintain both a predictable food supply and a greater variety of foodstuffs. Between 8000 BC and AD 1650 the world population rose to about 500 million.

This bar chart, on the right, plots the growth of the human population since the beginning of human life on earth.

Tokyo, the capital city of Japan, has expanded rapidly during this century. The map on the left shows the extent of growth since 1884, including land reclaimed from Tokyo Bay.

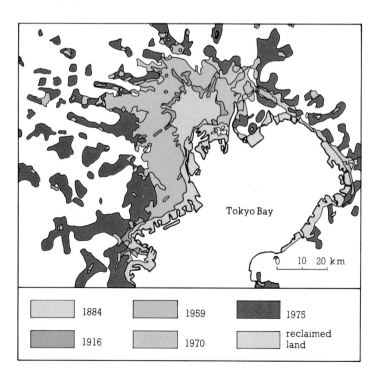

Tokyo Bay

1884 | 1959 | 1975
1916 | 1970 | reclaimed land

Bar chart values:
- 3 million — fire, toolmaking, up to 10 000 BC
- 5 million — agriculture, domestication of wild animals begins, 10 000 BC – 8000 BC
- 500 million — stable agriculture followed by trade, 8000 BC – AD 1650
- 4.4 billion — improved farming techniques followed by industrialization, 1650 – 1980
- 6 billion — predicted increase 1980 – 2000

The present expansion of the human population began with the improvement of farming techniques. Selective breeding produced hardier and more nutritious plants and animals, thus improving the diet. Cultivation was made easier and more efficient by the invention of such tools as the seed drill, which distributes seeds evenly in the ground. Population has also increased because of great advances in medicine that have reduced death rates.

The human population is now over 4000 million and is likely to rise to over 6000 million by the year 2000. Population increase, however, is uneven. The lowest rate of increase is in the western industrialized nations and the highest is in the developing nations, such as India.

It is not true, as is commonly thought, that there are too many people in the world already and that there is not enough food to feed them. There is, in fact, enough food for everyone but it is not evenly distributed over the world. In poor countries food is sometimes wasted because of inadequate storage facilities – the food rots, or rodents and insects eat it.

Poverty and starvation are also linked with the unequal distribution of the world's wealth. Some ecologists have speculated that the pattern of population growth can be explained by the theory of the ecological niche. Rich people have the broadest niche; but they produce few offspring because their needs as individuals are large. The needs of poor people are few. They produce more offspring because their requirements as individuals are small. They therefore have a narrow niche. Population growth is certainly lowest among the richest nations but, even within their class structure, patterns similar to that of the whole world can be observed. The richest classes, for example, control their numbers by social attitudes.

This theory may help to explain why human populations grow in the way that they do, and may also help us to understand better how to face the problems of population growth in the future. The inadequate distribution of food and resources is the main difficulty. If nations can find ways of overcoming this problem and ensure that aid reaches those people who really need it, then the future for the human race as a whole will improve.

Hong Kong is one of the most crowded cities in the world. Many of the poorer people live in junks and sampans on the waterways, such as Aberdeen Harbour, in this photograph.

The Control of Diseases

Humans and their domestic animals can be infected by many types of micro-organisms that cause disease. Some diseases, such as influenza and polio, are transmitted directly from an infected individual to a healthy one. Others, such as malaria and bubonic plague, are transmitted through other organisms, known as vectors. A genus of mosquito, called *Anopheles,* which carries malaria, is an example of a vector.

If a disease reaches a new area or host that it has not met before, it often becomes an epidemic, which means that it goes out of control. The influenza that broke out in Europe at the end of World War I spread rapidly around the world. As many as 20 million people died from the disease because no one had any immunity (natural resistance to disease) since it was a new type of influenza. Many populations have, however, acquired a natural immunity to local diseases. The people native to certain regions of Africa, for example, have a genetic immunity to malaria.

Despite occasional natural immunity, most diseases are a serious problem. Great efforts have been made, particularly by the World Health Organization, to wipe out various diseases. Scientists now think that smallpox, a disease caused by a virus that is transmitted directly between humans, has disappeared. With the exception of an accidental outbreak in a medical school in Britain, the last reported case of smallpox was in Somalia, Africa, in 1977.

This man, standing in marshy water, is spraying mosquito larvae with a pesticide to prevent malaria. This project against the disease was undertaken by the World Health Organization.

In the photograph on the left, researchers from Egypt's Institute for Medical Research are collecting snails in the Nile delta in order to study how the disease bilharzia is transmitted from snails to humans. Bilharzia has increased in Egypt because irrigation project have created large new habitats for the snails.

The blood-sucking tsetse fly shown in this photograph has landed on the fur of a water-buck in Africa.

Other diseases that are transmitted directly, such as diphtheria, can also be prevented by vaccination, the injection of a dead or living, but harmless, virus that produces immunity. Diseases carried by water, such as cholera, are often a problem in overcrowded slums, but can be controlled by better housing and improved sanitation.

Diseases carried by vectors are hard to control because it is the vector rather than the disease that must be stamped out. There have been some successful campaigns. Brazil has always had some malarial mosquitoes but, in 1929, African mosquitoes were introduced accidentally. By 1939 thousands of people were ill or dead. Close study showed differences in the behaviour of African and Brazilian mosquitoes that were the reason for the epidemic. African mosquitoes, unlike the Brazilian type that bred in forest ponds, were breeding in open, sunny pools and entering the houses where most of the infection started. The African mosquitoes were wiped out in three years by covering their breeding sites with oil. This prevents the larvae, which breathe air, from coming to the surface of water. The walls of the houses were also sprayed with insecticides. A similar campaign was launched in Africa but it has failed because the mosquito has become immune to most of the common insecticides.

Some diseases carried by vectors are spreading because of an increase in the habitat of the vector. In the Upper Nile populations of snails that carry bilharzia, a disease that attacks the liver, and blackflies that carry river blindness, which attacks the eyes, have increased because large areas of still water, which they inhabit, have been created by the Aswan and Sennar irrigation projects.

Controlling some diseases by eradicating the carrier may be impractical and people may have to adapt their farming methods instead. The tsetse fly, for example, inhabits damp areas in Africa. This insect transmits sleeping sickness, a disease that invades the nervous system, causing drowsiness in humans and in cattle. Removing this disease would mean cutting down large areas of forest. Zebras and some antelopes, unlike cattle, are immune to the disease. If the farmers can be persuaded to graze these animals, which are quite edible, instead of cattle, the likelihood of catching the disease will be greatly reduced.

Pollution

Pollution is a change in the physical, chemical and biological environment that can cause harm to humans and other animals and plants. Its effects are easy to see. When oil is spilled at sea or on the coastlines many birds and other sea creatures become coated with the oil and may die.

The major pollutants of many industrialized nations are the by-products of manufacturing industries. The waste gases given off by power stations that burn coal and oil, for example, contain sulphates. These are carried by the wind over the earth and fall in rain, as weak sulphuric acid, in places a long way from the source. Sometimes pollutants can be trapped above cities, such as Los Angeles, that lie in dips in the land. Other pollutants come from farming practices, including the use of insecticides and fungicides that are absorbed by plants and animals and are passed down food chains.

Animal and human sewage are often responsible for pollution in poorer countries. This type of pollution leads to the outbreak of diseases such as cholera and typhoid. Increasingly, however, these nations are also becoming polluted by industrial and mining wastes. In Malaysia, for example, copper mining, has polluted rivers and seas with low levels of copper, chrome and zinc, all of which are absorbed into food chains and are poisonous to higher animals, such as fish. Since the start of this mining operation in about 1977, sediments from strip mining, the removal of topsoil in order to reach the metals below, have been washed over rice fields and have ruined the crops every year.

Many pollutants, including the insecticide commonly known as DDT, remain in the environment for a long

Oil that is spilled at sea or near coasts causes the death of many organisms. This velvet scoter, a sea duck, has been caught in an oil slick and will die because its oil-coated feathers prevent it from flying and swimming.

The Tinto River in Spain is badly polluted by copper from nearby mines, and few fish or plants are able to live there.

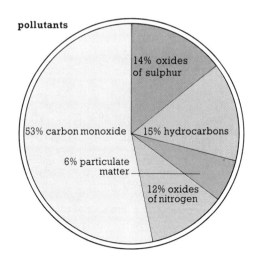
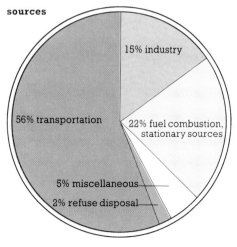

The diagram on the left shows the estimated percentages of various air pollutants in the United States in 1977 and diagram on the right shows the percentages of the sources of the pollutants.

Warm air normally rises and becomes cooler with height, as in the top diagram. In a temperature inversion, shown in the lower diagram, a layer of stagnant warm air, trapped by the cool air of the mountains, lies over the city like a blanket, preventing the polluted fumes of factories and cars from escaping into the atmosphere.

time because they do not break down naturally into harmless substances. The only effective control of such substances is a ban on their use and manufacture. Unfortunately, many nations cannot afford the expensive alternatives. Countries such as India continue to use dangerous insecticides for this reason despite the fact that many insect pests are now becoming resistant to them. It would, of course, be better to use chemicals that are selective (kill only a chosen species) and also to use those that can decompose naturally.

Sewage, although it is a pollutant, can be broken down but it nevertheless remains a problem because it is often discharged locally. Raw sewage contains millions of decomposer bacteria that require oxygen in order to function. When these bacteria enter rivers and seas in sewage they use up oxygen, causing the quantity of dissolved oxygen in the water to fall. This often makes the water unsuitable for fish and plants to live in.

Sensible laws and applied ecology have helped to remove pollutants from some areas. The Thames River, for example, was one of the most polluted rivers in Europe in the 1950s with no fish in its lower reaches. It is now one of the cleanest European rivers. Wastes from factories have been removed and raw sewage is no longer poured into the river, but treated to remove its oxygen demand by allowing it to decompose. In 1980 and 1981 there were reports that salmon had returned to the Thames – for the first time in over one hundred years. This encouraging example shows how applied ecology can overcome pollution problems.

The Problems of Intensive Agriculture

Modern farming produces as much food as possible from cultivated land for human and animal consumption. Unlike natural ecosystems, in which food is cycled between all members of the community, crops are managed for the sole benefit of human beings and their domesticated animals. Not only are other consumers prevented from eating any of the crop, but also the total yield of the crop is harvested.

In a natural ecosystem only a small proportion of plants are eaten and the rest is decomposed, freeing the nutrients that they contain into the soil so that they can be absorbed again by other plants. Our crops, such as sugar beet and sugar cane, are often completely cut down at harvesting and the parts that are not used are often burned. Little is available to the decomposers. As a result, there is no recycling of nutrients and the amount of humus in the soil is gradually reduced, leading to infertile soils.

In order to keep the yields of crops high, nutrients, such as nitrates and phosphates, have to be applied to the soil in the form of artificial fertilizers, made up into chemical compounds such as ammonium nitrate and potassium phosphate. Often too much of these fertilizers is applied. They may be washed from the soil because there is little humus to hold them in place. Fertilizers may then drain through the soil into rivers and lakes and eventually into drinking water. In water, fertilizers create eutrophic conditions, often resulting in spectacular blooms of algae. Nitrates in drinking water are a health hazard as they reduce the ability of the haemoglobin to carry oxygen in the blood of babies and children, leading to inactivity and sometimes death.

Intensive farming often uses powerful chemicals – insecticides, rodenticides and fungicides – to fight pests and diseases. Unwanted weeds are killed by applying herbicides. Unfortunately, most of these chemicals are passed on to other plants and animals in food chains. Some fish can be killed by only tiny quantities of herbicides and birds of prey have suffered similarly as a result of the use of insecticides.

Large amounts of fuels, such as coal and oil, are needed in order to produce these fertilizers and chemicals. In some areas of the United States almost as much energy is used to produce crops as the crop produces itself. Fertilizers and chemicals also cause pollution and in some cases, particularly where phosphates are mined from Pacific islands, they can destroy entire ecosystems.

The farming methods used in temperate regions are now being applied in the tropics and subtropics where the soil and climate are very different. These attempts at improving agriculture are often disastrous. Tropical soils have been cleared of their natural vegetation in

The demand for phosphate for use as a fertilizer has sometimes resulted in the devastation of natural habitats. For example, the removal of vast quantities of this mineral from Ocean Island in the South Pacific has completely destroyed the island.

These combine harvesters are working on wheat fields in Montana, USA, that were originally grasslands. The landscape now has few trees and almost all the plants and animals in the area are pests of cereal crops.

order to grow commercial crops such as cocoa and groundnuts. The fertility of these soils has rapidly fallen as heavy rain, no longer restrained by the thick, natural vegetation cover, takes away nutrients from the areas where the plant roots grow. Such soils have to be artificially improved with huge amounts of fertilizers. This practice is often uneconomical, as the cost of the fertilizers can be higher than the value of the crop.

Intensive farming has an uncertain future as the cost of fuels and their limited supply will start to restrict the use of chemicals and fertilizers. There are, however, practical alternatives based on ecological principles.

Crops can be moved around and the land can be restored in order to increase its fertility and reduce pests. Animal and human waste products can be applied to the soil instead of fertilizers. The energy needed by farm machinery can also be supplied by methane, the gas produced from the products of fermentation and digestion in plant and animal remains. Encouraging such new methods will greatly improve the environment without reducing the amount of food and resources that we need. Ecologists are keen for all nations, rich or poor, to adopt such ecologically sound agricultural practices.

Biological Pest Control

In many farming areas insect pests and weeds are a serious problem, especially when there are sudden increases in the sizes of their populations. Such outbreaks, caused by unstable systems of farming, such as monocultures, often result in serious crop damage. In many places the solution to this problem has been the use of insecticides and herbicides, but this often leads to the resistance of the pest to the applied chemical, and sometimes even to an increase in pest populations.

During the 1960s certain nettle caterpillars reached very high numbers on oil palms in Malaysia. The insecticides DDT, eldrin and dieldrin were used to control the caterpillars with good results at first. Within a few years, however, the populations of the caterpillars had increased to levels beyond those when spraying had started and, by the early 1970s, palm oil production had dropped more than forty per cent. The caterpillars had developed an immunity to the insecticides and their natural enemies had been killed by the insecticides. Natural enemies have now been brought in, the use of insecticides stopped, and the caterpillars are now controlled biologically.

All pests have natural enemies, such as parasites, predators, and micro-organisms that cause disease. These can can be used to control pests in an ecologically sound way, known as biological control. Biological control is not new. The ancient Chinese, about 1000 BC, fostered ants in their citrus trees to stop caterpillars and boring beetles from ruining their crop. This is a practice that is still continued in Burma and China.

During this century there have been some well-known biological pest control programmes. Prickly pear cactus in Australia has been reduced by a cactus moth. The walnut aphid, a pest of walnuts in California, has been controlled by using a parasitic wasp from Iran that attacks only the aphid.

Biological pest control does not completely kill off a pest species but keeps its population at a tolerable level. In greenhouses in Britain and The Netherlands, red spiders, a pest of cucumbers, are now controlled by predatory mites. To prevent damage to the mature cucumbers, the red spiders are introduced to the crop soon after it is planted. Then the predatory mite is also introduced, establishing a natural balance between the predator and its prey. Although the pest is still on the crop, the damage that it causes is minimal.

The photograph on the left shows dense prickly pear cactus in the Chinchilla area of Queensland, Australia, in 1926. The photograph on the right shows the same area in 1929, after the prickly pear had been controlled by the introduction of a cactus moth.

Apart from using natural enemies, biological control also involves the use of breeding techniques to increase resistance. The Hessian fly, a serious pest of wheat in the United States, is now much reduced because varieties of wheat have been bred to produce a chemical that is lethal to the parasitic fly.

Some pests can be avoided by carefully timing the planting and harvesting of the crop. Many insect pests will only attack crops at a particular stage of growth. In the USSR, for example, late planting of summer wheat reduces attacks by a flea beetle. In India, sowing wheat early avoids attacks by a gall moth. Alfalfa, a crop for feeding cattle, can be destroyed by lygus bugs. If alfalfa is left for a period of time natural enemies of the bug invade the area and attack the insect. Harvesting alfalfa by strip rotation also controls the bugs because their predators can move from cut strips to uncut strips, following the movements of their prey.

New developments in biological control include the release of sterile males into populations of pests. These infertile males prevent young from being born. Chemicals can also be used to lure males from females, and artificial hormones can stop pests from growing into adults. Both these methods considerably reduce pests. Such means of control provide a useful alternative to potentially dangerous chemical sprays.

This alfalfa in the United States is being cultivated in strips to prevent pests. Native to central Asia, this crop was introduced to the Americas by Spanish colonists. Prolific and high in protein, alfalfa also improves soil because it has nitrogen-fixing bacteria in its roots. Alfalfa is also grown in Argentina, southern Europe and Asia.

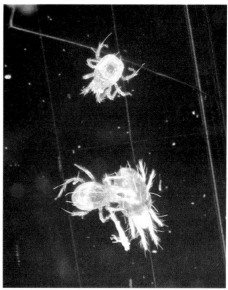

Taken through a microscope, the photograph above shows a red spider (it is actually grey in colour) and two of the predatory mites that can keep the pest's numbers in check.

The Great Experiment

The Earth's atmosphere is made up of about seventy-eight per cent nitrogen, almost twenty-one per cent oxygen and 0.03 per cent carbon dioxide. Other gases, such as sulphur dioxide and argon, make up the rest. These gases are cycled between the atmosphere, the oceans, the Earth's crust, and living things.

The proportions of the gases in the air have varied since the beginning of life on Earth. When life began, between three and four thousand million years ago, the atmosphere contained gases produced by volcanic eruptions. There was less oxygen and more carbon dioxide than now. As living forms, such as algae, evolved, they altered the atmosphere by releasing oxygen into it. About 600 million years ago the level of oxygen may have reached three per cent.

This small quantity of oxygen gave creatures such as sponges, corals, and shellfish the opportunity to increase their numbers. As they grew, they took in a huge amount of carbon dioxide that was absorbed into their bodies to form calcium carbonate. The skeletons of these animals are the limestone, dolomites and chalks that are found in rocks and cliffs around the world today.

The present quantities of atmospheric gases were reached about 350 million years ago. Since then, there have been further alterations in the relative quantities of gases in the air, particularly between carbon dioxide and oxygen. About 250 million years ago there was a rise in the amount of carbon dioxide, and the climate of the Earth warmed up. This may have been caused by an increase in volcanic activity. At this time the remains of vast numbers of green plants were being turned into peat because they did not decompose. As the layers of peat sank down, the pressures in the Earth's crust pressed the peat layers together to form coal. As a result, the quantity of carbon dioxide in the atmosphere may have fallen to its present level because the carbon was incorporated into the peat and not returned to the atmosphere. Since then, there have been only small changes in the quantities of atmospheric gases.

Carbon dioxide is cycled between the air, plants and animals, and between the air, oceans and the Earth's crust. The oceans act as a balance, dissolving a certain amount of gas from the air and releasing it when atmospheric levels fall. Agriculture and industry are, however, causing a rise in the quantities of carbon dioxide in the air. The vast deposits of carbon in coal and oil are being burned for fuel, adding to the carbon dioxide in the air. Farming methods that speed up the decomposition of humus, such as turning over the soil and forest clearance, are also adding carbon dioxide to the air.

Vast quantities of carbon dioxide are now being released, and the amount is steadily going up. The oceans, for unknown reasons, are not absorbing enough carbon dioxide to compensate for the altered

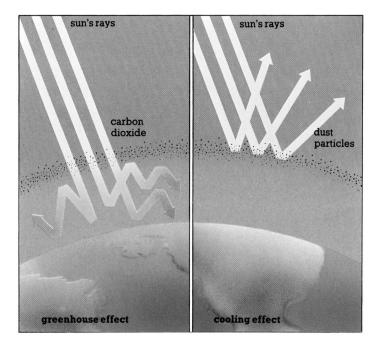

This diagram illustrates how the Earth's climate may be changed either by increased carbon dioxide being released into the atmosphere or by dust particles. On the left, the greenhouse effect is shown, with the extra carbon dioxide letting the Sun's rays through but also absorbing infra-red heat, so causing the Earth to become warmer. On the right the cooling effect is shown, with dust produced by burning fuels reflecting the Sun's rays, so keeping the Earth cool.

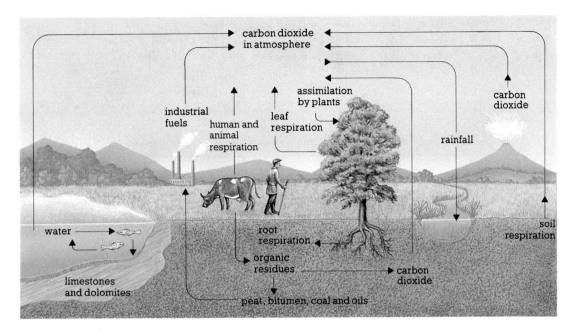

Carbon, like other gases and vapours, is recycled through plants, animals and the soil. This illustration shows not only the natural carbon cycle but also how increased carbon dioxide is being contributed to the atmosphere, and joins the natural cycle, by the burning of peat, coal and oil.

levels of the gas. Many scientists think that by the year 2000 the amount of carbon dioxide in the air will have risen by about twenty-five per cent and could reach twice its present level by the mid twenty-first century.

Some scientists believe that this increase will cause a warming of the Earth, called a greenhouse effect. Carbon dioxide is transparent to the Sun's rays but, like glass, absorbs infra-red heat. The Earth will therefore warm up because the infra-red heat will be unable to escape from the atmosphere. If this actually happens it is possible that the polar ice caps will eventually melt, causing a rise of about sixty metres in sea level. This rise in water would drown coastal cities and low-lying areas all over the world.

Burning fuels, however, also releases dust into the upper layers of the air, which reflects the heat of the Sun's rays. Some scientists think that the cooling produced by this reflection will be enough to prevent the greenhouse effect. Such reflection might even bring down the temperature of the Earth by 3°C – enough to cause another Ice Age.

Coal-fired power stations produce sulphates, nitrates and fine dust particles, which often pollute surrounding areas. A tall chimney, such as this one at a factory in West Germany, reduces local pollution but releases pollutants higher into the atmosphere, where they circulate in air currents and may fall in areas remote from their source.

Food for the Future

About 700 million people, mainly in Asia and Africa, are estimated to be in extreme poverty, many of them starving or lacking the minerals, vitamins and proteins that are essential for good health. Yet enough food is produced to feed the total population of the world. The problem lies instead in the types of food that are produced and the way in which they are distributed.

In recent years, the world has produced about 1 300 million tonnes of grain each year. The developed countries, such as the United States and those of Europe, consume half this amount although they represent only one quarter of the world's population. Much of this food is used to rear livestock rather than to feed humans. Farm animals consume two-thirds of the world's grain harvest, some of which is imported into developed countries from underdeveloped ones where the people are short of food.

Intensive animal rearing is an inefficient way of producing protein, the complex elements that are an essential food item. Seven kilogrammes of grain and soyabean, for example, are needed to produce only 0.4 kilogrammes of beef, and three kilogrammes are needed to produce 0.4 kilogrammes of pork. These cereals and legumes could be used directly to feed people, since the correct mixtures of plant proteins provide a good diet. These basic foodstuffs could also be grown in poor countries in place of 'cash crops' such as tobacco and coffee, which are exported to wealthy countries.

Ecologists are also in favour of rearing cattle and other animals on poorer land. The quality of the meat would not be greatly impaired, and the more fertile land could be used for growing crops to eat instead. Much waste land could also be made fertile by recycling manure back on to the land.

Attempts have been made this century to increase food production in underdeveloped countries by carefully breeding plant crops that have a high yield. This was called the green revolution and was successful in increasing yields of grain in Mexico during the 1940s. Unfortunately, such methods can have harmful side effects, as they did in the Philippines during the 1960s. New varieties of rice were grown that produced high yields, but they were badly attacked by pests, needed large quantities of fertilizers, and were lower in protein than the old varieties of rice. The price of seeds, fertilizers and pesticides that were required was too high for many of the farmers. Furthermore, fishes in the nearby rivers were poisoned by the fertilizers and pesticides, thus reducing another important local source of protein.

None of these world food problems are insoluble. If

This rice-threshing machine is being used in Thailand, where, formerly, rice would have been threshed by hand. It is easy to maintain and repair, but unfortunately has to be powered by a diesel engine that runs on increasingly expensive fuel.

In this field in Java two crops are being grown together. One is pineapple and the other is a newly introduced variety of wing bean, which is the most nutritious food crop in the world. All the plant's parts, which are high in protein, can be eaten by humans, except the stalk, which can be fed to animals. Wing bean roots also contain nitrogen-fixing bacteria that help keep the soil fertile. Both crops can be eaten by the local population or exported.

Ecologists are in favour of rearing cattle and other animals on poorer land, such as this hillside in the Algarve, Portugal. The quality of the meat is not affected, and the more fertile land can be used for growing crops instead. Waste land can also be made fertile by recycling manure back on to the land.

different crops were grown at different times, those that drain the soil of nutrients can be alternated with legumes, which are high in protein, and which restore soil fertility. Shellfish, which are also a good source of protein, can be cultured in shallow, coastal waters that are rich in nutrients. More plant protein could be used rather than animal protein. There are about 80 000 known edible plant species yet only 200 of them are used, at the moment, to provide us with food.

Research into local varieties of plants as food sources might help underdeveloped countries more than importing expensive Western technology and crops. Such technology, however – if it is used in the right way – can also greatly improve the development of new food sources. *Spirulina,* a species of freshwater algae, has a very high protein content and scientists believe that its yield may be ten times greater than that of soya. It is now being grown on a commercial scale.

International aid for underdeveloped countries can also be directed at improving food storage. The places used for storing grain in Asia, for example, are not very good and much of the grain is eaten by insects and rodents or rots away in warm, tropical climates. Concrete and metal silos would solve this problem. Similarly, better roads, refrigerated lorries and railway trucks would reduce the rapid decomposition of food.

Recycling Resources

Natural resources, such as metals, fuel, minerals and wood, which we need for our factories, farms and for everyday use are limited in quantity. By the turn of the century it is possible that the known reserves of many ores will have been used up. Copper, for example, may be nearly exhausted while others, such as cobalt and titanium, will be so scarce that they will be very highly priced. Increasing demands for wood are also removing forests faster than they can grow.

Mining, and manufacturing metals into finished goods, uses up fossil fuels, such as coal and oil, resources that are also limited in quantity. Pollution is created when the metals are mined, and in the places where they are made into products. These products, such as cans and bottles, pollute the environment when people throw them away. Manure is also thrown away, often into lakes, rivers and the sea, producing pollution in water.

Natural environments do not pollute themselves with waste products: minerals and nutrients are recycled. Today we are beginning to realize that we need to copy how nature works – in other words, to learn to recycle our resources in the same way. Recycling both conserves materials for further use and reduces the amount of pollution.

Many metals, because of their value, are already being recycled. Old car batteries, for example, are salvaged for their zinc. Aluminium containers are easy to recycle. Tin cans, which are actually made of aluminium and steel, can be processed to remove the lead. The steel left over from this process can be re-used in the steel manufacturing industry.

Glass can be recycled in various ways. Glass bottles can be returned to factories to be washed and refilled. Many cities, including London, now have 'bottle banks' where people can deposit old bottles for re-use. Britain also has a long tradition of using returnable glass bottles for milk. In the United States, eight states have banned the sale of non-returnable drink containers. Glass can also be broken up and used again.

Most of the timber felled today is used for pulping into paper. Most of the world's demand for paper could, in fact, be met by recycling. Good quality writing paper, newsprint, paper bags and cardboard can all be made from used paper that would otherwise be thrown away.

Plastic containers are particularly bad for the environment because, when they are discarded as litter, they do not decompose. Now, however, rubbish containing plastic and paper can be processed to produce a fuel suitable for use in industry. Used plastic bottles can also be shredded to make insulation for sleeping bags, or spun into fibres for synthetic cloth.

Manure can be used as natural fertilizer on farmland and to help reclaim derelict land where the soil is infertile. Manure can also be processed to produce methane, the colourless, odourless gas that can be used as a fuel both for industry and for domestic use.

Recycling the things that we use in our everyday life – metal from cars, paper, bottles and refuse, among other examples – has exciting possibilities for the future. It is just one of the ideas encouraged by ecologists and shows how ecological principles can be applied in a practical way to benefit our industry and agriculture as well as our environment.

This recycling plant in Holland is the most modern in Europe. Waste is brought to the plant by rail and left in heaps so that the organic material can rot. Then the waste is divided up into plastics, compost, paper and metal (glass is not recycled in this plant). Waste paper, which is being processed in the photograph on the right, is recycled as board and newsprint.

Glossary

Adaptation: a characteristic of organisms that improves their chances of survival in the environment in which they live

Algae: a major group of plants of very simple form that are found in water and damp places

Bacteria: a group of microscopic organisms found in soil and water, or parasitizing other organisms

Biome: a major community of plants and animals representing the climax of a climatic region

Carnivore: an animal or plant that feeds on flesh

Chlorophyll: the green colouring matter found in plants. It is used in the process of photosynthesis.

Climatic climax: a type of community that can continue to live and reproduce indefinitely in constant climatic conditions

Community: a group that consists of populations of plants and animals that intereact with each other

Consumer: the name given to an organism that eats other organisms, either plants or animals, or their remains

Decomposer: an organism that breaks down the bodies of dead plants and animals into simple substances

Detritus food chain: a food chain based on decaying organic matter

Dominant: a plant or animal species that controls the conditions and character of a community because of its numbers, its size, the area that it covers or by its activities

Ecological pyramid: a pyramid-shaped graph that shows the mass, numbers, or energy levels of the individuals of each feeding level in an ecosystem, starting with the primary producers at the base

Ecosystem: a community of organisms that relate to each other and also interact with their non-living environment

Edaphic climax: a climax community that differs from the climatic climax community of the area where it is found because of local differences in soil conditions, such as a hardpan next to sandy soil

Environment: all the living and non-living factors that influence an organism

Evolution: the changes in the characteristics of organisms that take place through succeeding generations, often over thousands or millions of years

Food chain: a chain of organisms through which energy is passed. At each step in the chain some energy is lost, restricting the length of the chain to four or five different types of feeders.

Fungi: a major group of organisms that live as parasites, saprophytes, or symbionts

Gause's principle: a principle that states no two species can occupy the same niche in the same place at the same time

Grazing food chain: a food chain that is based on the material produced by photosynthetic plants

Habitat: a place with a particular environment that is inhabited by organisms

Herbivore: an organism that eats plants

Humus: complex organic matter found in soils that results from the decomposition of plant and animal remains

Legume: a member of a large group of plants with nodules on their roots that contain bacteria capable of converting nitrogen from the air into soluble nitrates

Mycorrhiza: symbiotic relationship between a fungus and the roots of plants

Natural selection: the natural process whereby the organisms that are best adapted to their environment survive while those that are poorly adapted do not survive

Niche: the role of an organism in its environment together with its activities and relationships with other organisms

Omnivore: an animal that eats both plants and animals

Organism: anything capable of life processes

Parasite: an organism that feeds on another living organism

Photosynthesis: the process by which plants use the energy of the sun to convert water and carbon dioxide into sugars, thereby transforming light energy into chemical energy

Phytoplankton: microscopic green plants that float or swim in water

Pioneer community: the community that first colonizes barren soil

Plankton: floating or swimming microscopic plants and animals found in water

Pollution: contamination of an area by substances that make the area unsuitable for most organisms to live in it

Population: a group of organisms of the same species that interact with each other

Predator: an animal that kills others for food

Primary producer: an organism that uses sunlight to make food from simple substances

Productivity: the total quantity of organic material produced within a given period by the organisms of a system, or the amount of energy contained within the produced material

Recycling: passing through a system in order to be used again

Respiration: the breakdown of sugars to release energy

Saprophyte: an organism that feeds from the remains of dead or decaying plants and animals

Scavenger: an animal that eats the remains of animals killed by other organisms

Species: a group of organisms that can breed with each other but not with other groups of organisms

Succession: progressive changes that cause one community to be replaced by another

Symbiosis: two or more organisms living in close association for their mutual advantage

Territory: the area occupied by an individual or group of organisms that is defended from invasion by other organisms, usually of the same species

Trophic level: a part of a food chain in which all the organisms obtain their food in the same way

Zooplankton: minute floating and swimming animals that feed from microscopic green plants or other planktonic animals

Index

adaptation principle 65
agricultural problems 84-5
air pollutants 83, 88-9
algae 8, 10, 15, 17
    freshwater 42
    marine 45, 46, 49
alpine environments 53
animals 28, 29, 80-1
    desert regions 54-5, 74-5
    forests 59, 61, 62-3
    grasslands 56, 57
    marine ecosystems 44-50
autecology 6

bacteria 8, 10, 17, 43, 49
balance of nature 5
biological control 86-7
biomass 33
biomes 40-1

carbon cycle 17, 88-9
climate 5, 40-1, 54
    climax communities 26
    deforestation, effect 68-9
    soils 12
coexistence 20, 21, 36-9
colonies 8
colonization, land 24-5
communities 22-3, 32-3
    climax 25, 26-7
    ecosystems 8, 9, 28-31
    grasslands 56
    rain forest layers 62-3
    succession 24-5
competition 36-7
coral reefs 48-9
coniferous forests 58-61
    desert soils 75, 76
    forests 68-9
    marine life 70-1
    soils 66-7
    waste products 85, 92-3
    wildlife 72-3
consumers 28, 29, 30-3
cyclic systems
    climatic 40
    climax communities 27
    energy 14-15, 16-17
    nutrients 16-17, 88-9
    populations 34-5

Darwin, Charles 64-5
deciduous forests 58-61
decomposers 10, 11, 28, 29, 30
deforestation 68-9
deserts 54-5, 74-7
detritus feeders 30, 49
diseases, control 80-1
dominance 22-3

ecoclines 40
ecological niche 20-1, 79
ecosystems 6, 28-9, 73
    coastal regions 46-7, 50-1
    deserts 54-5
    energy patterns 28, 30-1
    forests 58-63
    freshwater 42-3
    grasslands 56-7
    marine 44-51
    tundra 52-3
edaphic communities 26
environment 6-7, 24-5
    man's influence 64-5
    pollution 82-5
erosion 66-7, 69, 71
estuary habitats 46-7
evergreen trees 60-1
evolution theory 64-5
extinction effect 71, 72, 74

farming
    deserts 74, 76-7
    soil erosion 66-7, 74
fertility, soils 11, 12, 84-5, 90-1
fertilizers 19, 84
fish industries 70-1
food chains 30, 32-3
    coral reefs 49
    herrings 71
    mangrove swamps 50
    marine 44
    food supplies 15, 18, 19, 79, 90-1
forests 58-63
    communities 26-7
    management 68-9
freshwater ecosystems 42-3

Gause, G.F. 36
genetic engineering 72
grasses 46, 47, 57
grasslands 56-7
greenhouse effect 88-9

habitats 20-1, 22-3, 72-3
horizons, soil 11
human populations 78-9
humus, soil 10, 11, 61

insecticides 7, 82, 84, 86
irrigation, deserts 76-7
individuals 9

lake environments 42-3

Malthus, Thomas 64
mangrove swamps 50-1
marine ecosystems 44-51, 70-1
minerals, soil 10, 16
natural selection 64-5

nitrogen cycle 17
nutrient cycles 16-17

oases 76
organisms, ecosystems 8-9
    freshwater 42-3
    marine 44-51
    soil 10-11
overgrazing damage 67, 74-5
oxygen cycle 17

parasitism 38
pesticide problems 84, 86
pests, control 82, 86-7
people and ecology 64-5
photosynthesis 14-15, 18
phytoplankton 44, 45, 46, 49
plant energy 14-15, 18-19
pollinators 6-7, 52-3
pollution 82-3, 88-9
population 8-9, 64
    communities 22-3
    human explosion 78-9
    regulators 34-5
productivity pyramids 33

rain forests 58, 62-3, 68-9
rainfall 40, 54, 74-5
recycling methods 92-3
respiration, plants 14

saline soils 12, 77
savanna 27, 56
sea ecosystems 44-51, 70-1
seaweeds 45, 46
soil environment 10-13
    conservation 66-7, 69
succession sequences 24-5
sunlight 14-15, 18, 61
survival mechanisms 38-9, 53
symbiosis 39
synecology 6
systems, organization 8-9

taiga zone forests 60
territory principle 34
trophic levels 30-1
tropical regions 40, 41
    soil erosion 67
tundra ecosystems 52-3

vegetation zones 40-1
    deserts 54
    tundra 52, 53

water cycle 16
water ecosystems 42-51
wildlife conservation 72

zones, vegetation 40-1
zooplankton 45, 49
Zooxanthellae 49

Credits

The publishers gratefully acknowledge permission to reproduce the following illustrations: Adespoton 77t; H. Angel 19, 24, 36, 43b, 46r, 47, 68, 91b; Aquila 33; Ardea 23, 57, 66, 69t, 82; Australian Information Service, London 86l, r; Bavaria Verlag 59r, 77b; Biofotos 15t; Biophotos 71; F. Bruemmer 563t; P. Buringh, *Introduction to the Study of Soils in Tropical and Subtropical Regions,* pudoc, 1968 10; J. Allen Cash 11; B. Coleman 3, 6, 7, 13, 37, 38t, 41, 46l, 51l, 73t,b, 74, 75t, 79, 81; Explorer 75b; Glasshouse Crops Research Institute 87r; R. Harding 53b, 63, 65b; E. & D. Hoskings 20b, 22, 70t, 83; A. Hutchison 65t; Jacana 38b, 61; F. Lane 12, 35, 59l; S. McCutcheon 52; G. Mazza 8, 49, 50; M.I.S.S. 64; T. Morrison 15b, 55b; Natural Science Photos 20t, 69b; Nature Photographers 32; Naturfotograferna Bildbyra 34; A. Notholt 84; J. Olmstead 26, 26b; Oxford Scientific Films 9, 27t, 29, 62; A. Ronan 2; J. Ross 90, 91t; Seaphot 43t, 45t, 48l,r, 51r; V.A.M. 93; Vision International 60, 70b, 78; W.H.O. 80l, 4; J. Frederick Grassle/Woods Hole Oceanographic Institution 45b; ZEFA 18, 55t, 67, 76, 85, 87l, 89.

Front jacket photograph: Premaphotos Wildlife
back jacket: Bruce Coleman

Artwork by: Gerard Browne 30-31b, 83b; Carol McCleeve 14r, 32, 35, 36, 41, 47, 83t; Richard Phipps 8-9, 11, 13, 16-17, 29, 56, 61, 63, 66; John Yates 14l, 21, 23, 28, 31 tl, 31 tr, 32, 35, 39, 40, 42, 44, 49, 51, 52, 54, 57, 71

Picture Research: Mary Fane

The publishers would also like to thank Sheila Silcock and Denis Filer for their especial help in preparing this book.